The Value of Species

The Value of
Species

Edward L. McCord

Yale UNIVERSITY PRESS

NEW HAVEN & LONDON

Published with assistance from the foundation established in memory
of Amasa Stone Mather of the Class of 1907, Yale College.

Yale University Press books may be purchased in quantity for
educational, business, or promotional use. For information,
please e-mail sales.press@yale.edu (U.S. office) or
sales@yaleup.co.uk (U.K. office).

Set in Minion type by Keystone Typesetting, Inc.
Printed in the United States of America.

Library of Congress Cataloging-in-Publication Data
McCord, Edward LeRoy.
The value of species / Edward L. McCord.
p. cm.
Includes bibliographical references (p.) and index.
ISBN 978-0-300-17657-5 (alk. paper)
1. Environmental ethics. 2. Species. 3. Biodiversity. 4. Extinction (Biology)
5. Nature—Effect of human beings on. 6. Environmental policy. I. Title.
GE42.M378 2012
179′.1—dc23   2011035192

A catalogue record for this book is available from the British Library.

This paper meets the requirements of ANSI/NISO Z39.48–1992
(Permanence of Paper).

10   9   8   7   6   5   4   3   2   1

To my mother and father, Betty and Guyte,
and to my grandmother Elizabeth Reed Mack

The World Is Too Much with Us

The world is too much with us; late and soon,

    Getting and spending, we lay waste our powers:

    Little we see in Nature that is ours;

We have given our hearts away, a sordid boon!

The Sea that bares her bosom to the moon;

    The winds that will be howling at all hours,

    And are up-gathered now like sleeping flowers;

For this, for everything, we are out of tune;

It moves us not.—Great God! I'd rather be

    A Pagan suckled in a creed outworn;

So might I, standing on this pleasant lea,

    Have glimpses that would make me less forlorn;

Have sight of Proteus rising from the sea;

    Or hear old Triton blow his wreathed horn.

—William Wordsworth (1806)

CONTENTS

PREFACE

D id you notice when the natural world first grabbed your attention? When you realized with all your sensibility that we are not alone—that our world teems with birds, bugs, reptiles, and other creatures that are certainly not human but are certainly also not unrelated to us? It is a stunning revelation and a memorable moment in a human life. It begins a journey that leads to frontiers of self-discovery and endless interplay between the life within us and the living world without.

We come from diverse homelands and childhoods that afford us varying engagements with the natural world. For my part,

I grew up in north Florida, and I lost myself with friends in outdoor adventures as a boy. Ours was a land of salt marshes and fiddler crabs, of black lagoons and dark ravines, of watery sinkholes with walls of limestone and ferns, of crystalline rivers winding through glades of cypress and Spanish moss, of the sounds of cicadas, owls, frogs, and whippoorwills, and gentle winds blowing through the pines, crashing claps of thunder, drenching warm rains.

Nature and the outdoors were a passion for my parents, who generously indulged the same in my brother, my sister, and me. With my brother's deft interceptions, we took in opossums, crows, flying squirrels, snakes, salamanders, raptors, turtles, and even a young green heron. I was captivated by everything great and small that crawled or flew or unfolded in growth. But my first revelation of the astonishing and improbable mysteries of nature among which we live occurred when I was eleven, when I opened a *National Geographic* feature on the carnivorous plants native to my area. The world was never the same for me after that. I knew then that we move among miracles.

North Florida is known for its wealth of carnivorous plants. In sandy coastal wetlands their sculptural leaves of red, yellow, and green glow in the sunlight across shimmering grassy landscapes of longleaf pine, sweet bay magnolia, and bald cypress. Exquisite terrestrial orchids grow in these same pine barrens, and they became another great passion of mine. *Calapogon tuberosus,*

Platanthera ciliaris, Cleistes divaricata: I pursued orchids in summer excursions from deep in the Everglades to the forests of the Appalachian Mountains and the quaking sphagnum bogs of northern Wisconsin, where I worked in a camp for the National Audubon Society.

When I was fourteen I read a passage in a horticultural encyclopedia that ignited my imagination as nothing had before. It described a towering mountain on the island of Borneo, a mist-shrouded pinnacle that harbored fantastic species of carnivorous plants found nowhere in the world but on its steaming slopes along with a thousand species of orchids. After years of dreaming about the trek, in the summer after my first year of law school I hiked up Mount Kinabalu, reaching the summit at dawn. I was so engrossed and excited that I lost all track of time.

Children everywhere awaken at an early age to countless lifelong interests—airplanes, horses, computers, baseball. Who knows how or why this happens? They awaken as well to awareness of other life, each in his or her own way. Every child at some time surrenders, even if fleetingly, to the fascination of a living thing—a sparrow, a tadpole, an ant, a porcupine, a mushroom. Humans surely have roused to the wonders of existence since time immemorial. But there is a new wind blowing among youth today. They are attracted in great numbers to courses in "environmental studies" that were completely unknown a few decades ago.

I have been teaching these courses for a dozen years now. It is something I had never planned to do when I was in graduate school studying the philosophy of science and theories of meaning, or during my years as a lawyer. But I found the opportunity that was offered to me by my university irresistible, for I have come to believe that environmental studies deal with the most profound and urgent frontier that humanity has ever faced.

I wrote this book on a guest ranch in Wyoming, where I am the director of a multidisciplinary Yellowstone Field Course for the Honors College at the University of Pittsburgh. Our classroom at the site of the world's largest volcanic caldera is only the grand sky above us and the ancient processes of geology and ecology surrounding us. I also oversee a 6,000-acre preserve for ecology, archaeology, and paleontology in the opposite corner of Wyoming. A generous cattle rancher donated this land to our university as an educational and scientific preserve. In 2007 our undergraduates, in their first project there, unearthed from a grassy hillside the shoulder blade of an *Apatosaurus*. To hold in one's hands the remnant of such a spectacular creature that roamed the earth millions of years ago is an extraordinary experience. But more wondrous still is to realize that we too are forms of life that evolved on the same planet as that dinosaur, and that we are standing in this moment of time and perceiving with a special intelligence the dinosaur's fossil remnant. Ours is a comprehension and sense of wonder unmatched, to the best of our knowledge, by any other species in the universe.

These reflections about human self-awareness and the associated values of discovery humans share, disconcerting in the practical world of affairs and largely neglected in the classroom, are the subject of this book. Environmental studies are not the discrete and separate topics that appear in course catalogues but an integrated web of interactive themes across the sciences and humanities. How could this be otherwise? These studies are surely ultimately and most relevantly about defining our human identity on earth. And each of us is a "person," not an amalgam of dissociated parts. Environmental studies must be routes to understanding, reconciling, and advancing our human values. In other words, to fail to see how a structure of values integrates the disparate disciplines of environmental studies is a failure to function as a healthy, effective, and whole person.

I offer this plainspoken examination of the value of species as a step toward that more clear and integrated understanding and to foster new perspectives upon the fate of life on earth and our priorities and impulses that relate to it. My aim is to provide a fresh, rigorous, and concise analysis of the human predicament with regard to protecting the heritage of the earth, a predicament marked by the fact that our species is contributing to the extinction of so many others.

The idea for this book arose in part from problems encountered in teaching environmental ethics and public policy. My students had no common background in philosophy, psychology, law, or the other disciplines essential to a thorough comprehension of what we were talking about each day. I realized that they needed a book that synthesized these fundamentals and integrated them into real-life challenges of environmental reasoning. I had served as co-counsel for a large real estate class action, during which I developed a practical sense of property rights law. As a Fulbright-Hays Fellow in Mongolia, I had seen how property rights served the market economy among people who had never owned property before. These experiences enabled me to structure this book in a way that would supply the essential foundation for both informed general discussion and supplemental in-depth study.

I had another, more vital purpose in writing this book, one that gave me the energy and focus to pull it all together. Species vanish daily from our earth, and it is vital that we understand why we should care about these losses. The question that needed to be answered was "Do all species have a value to the public interest regardless of their individual use?" However steep the challenge of answering this difficult question, I began to see it more and more as a way to motivate a much-needed public awakening about our relationship with the natural world, and to rediscover my own bearings.

When I began teaching environmental studies I had been speculating for a long time on my youthful encounters with the environment. I came of age during a period of massive environmental transition that changed my early infatuation to a response that was more reflective and skeptical. Our values seemed to have become confused and more vexing than I had ever imagined they could be. The world that so enthralled me when I was young had become a very different world, covered by a dark cloud and in danger of destruction from the many onslaughts inflicted on it.

This book is a quest for some honest self-understanding and the will to stand against this sea of losses in our world. Would it not be an even greater order of human tragedy if we never recognized why it matters that so many species are vanishing forever? The question of why we should care is therefore central in these pages, and the answer goes beyond our functional needs. Our response should not primarily be conditioned by whether rain forest plants can cure cancer or other practical matters, however significant these may be. Our response should be grounded in the values of the spirit, anchored in the essence of what it means to be human.

This book does not, therefore, offer solutions to particular environmental challenges. In the face of the mounting decimation of species today, we need to revisit and clarify more thoroughly and deeply what the challenges are, and what our values should be.

There are many paths to a destination. In writing this book I am not suggesting that other scholars have not wrestled with some of the same issues I do. Although pieces of this story are being told within a variety of disciplines, however, they are not being told as part of a single narrative that integrates the challenges they present succinctly enough for readers to comprehend their unity. I therefore designed this examination to be as concise and explicit as possible so that its wide range of subjects could be comprehended within a single framework. My objective here is not scholarly exegesis but rigorous thinking stripped to fundamental observations and communicated plainly to a reflective, curious public. Further, unlike much scholarship, whose aim is a distanced objectivity, the advocacy in this book is deliberately passionate, and it must be judged by that passion in light of the epic proportions of the tragedy that is before us.

You may say, "The task of saving the life of our planet seems impossibly difficult." But this need not be the case. The factors in our nature that have brought civilization to this extraordinary predicament are not our fault—at least, they have not been thus far—and that is reassuring. There may still be time for us to make the behavior and attitude changes necessary to change our destructive course. Indeed, until recent times, we could not have known what we were doing, and we were propelled into this state of affairs by inexorable forces of consumption that govern all life on earth. We should also find solace in the fact that this, our human dilemma, is so fascinating in itself, revealing to us truths about our condition that no other generation in the history of humankind has previously had the privilege of understanding. But the burden is correspondingly heavy.

Further in response to those who believe that this task is too difficult, I would answer that such an attitude is surprising. For the task really is so simple. As simple as drawing a line in the sand, something we do all the time to protect important values. It is as simple as guarding those lines in the sand with all the force of our community, our heart and our soul, and our knowledge and resolve. This, again, is something we do all the time for our important values. If we cannot do it for something so vital as the life of the planet, the consequences for all our future generations are dire. In the final analysis, there are simply no costs of taking action that remotely compare with the cost of failing to do so.

In essence, if we are going to save the life on this planet for generations to come, we must begin to make ourselves whole and ask the fundamental question "What kind of humanity do we embrace?" We can argue forever about strategies. But if we cannot clearly and finally understand the values that are at risk and embrace them, then no strategies on earth can save us.

With all of our differences, we are still voyagers on this ship together. Imagine our moments of panic when we hear the captain's call, "Iceberg ahead!" Where will we be? Who will be with us? Will all the money in the world save us? And what does it mean to be saved?

Well, the last question is the important one. And the captain is calling.

ACKNOWLEDGMENTS

I came to the range of perspectives brought to bear on this book from experiences over much of my life, but one person in recent years directed opportunities my way that filled out the substance and pulled everything together. This was my dear friend, the late Alec Stewart, who built the University of Pittsburgh Honors College from its inception as an Honors Program in 1979 into the inspiration of intellectual attainment that it is today. The birth of this book stemmed directly from a conversation with Honors College students in the spring of 2006, and Alec gave me unflagging support as it developed. In this way, I share with all the students and staff who passed through Alec's life that sense of boundless debt to his gentle and deep influence. For me, the debt is rendered particularly poignant by the fact that Alec did not survive to see this book's publication. It is a lesson about life of which no one would have appreciated the wry humor more than him. This book is a testament to my measureless good fortune in having known Alec Stewart.

Others have stood by in these years with invaluable counsel. Foremost among them is my great friend Gary Lawlor, a prolific poet and voracious student of literature, whose gifted aesthetic sensibilities helped me find and re-find my voice through innumerable stumbles. Wyoming ecologist Gary Beauvais, instructor in the University of Pittsburgh Yellowstone Field Course, received the first raw pages of the manuscript day by day at his cabin next door to mine at the marvelous K Bar Z Guest Ranch. He understood what it was about, urged it on, and brought to it his sharp critical judgment and seasoned perspective. Paleontologist Bud Rollins invited me to teach with him the foundational class for Pitt's new environmental studies major, setting in motion the reflections that would form the basis of this book. He inspired me through our shared affection for the southeastern U.S. coastline and was the source of much support and scholarly counsel. Historian Neal Galpern learned about the manuscript one day and asked to see it, returning several days later with his copy meticulously annotated with incisive comments. It was a process of close reading that continued in critical stages, helping me surmount some fundamental hurdles in developing the work. Philosopher Nicholas Rescher kindly met with me through successive versions of the manuscript and offered his counsel and assistance at critical junctures. Philosopher John Earman had invited me to speak to his class on environmental philosophy years ago, and generously shared his insights when he reviewed the manuscript. My co-instructor in "Environmental Science, Ethics and Public Policy," Charlie Jones, a geologist and gifted teacher, brought his seasoned editorial and scholarly counsel to the manuscript repeatedly as we used successive versions as our class text. My good friend Ellen York from biological sciences shared her careful insights and editorial thoughts about the manuscript from the beginning. Philosopher of religion Nathan Hilberg of the Honors College shared crucial wisdom and advice from the book's inception. Colleagues Michael Giazzoni and Mandela Lyon, with specialties in higher education and paleontology, respectively, were generous with support and important editorial counsel through the years.

There has been no voice more profound in inspiration and critical in impact upon the fortunes of this effort than that of population biologist Paul Ehrlich, who generously brought his vital counsel and encouragement to this project and opened doors. Ehrlich introduced me to a masterful editor, Jonathan Cobb, to whom I owe a great sweep of indispensable advice and my gratitude for sharing regularly with me books by kindred authors. Editor Sarah Flynn also vouched for the manuscript at first read in a much earlier form, and with forthright counsel propelled me to write and then rewrite the Preface until I got it right.

At Yale University Press, I've been greatly privileged to work with Jean Thomson Black, who managed the editorial process with supreme skill. I'm especially indebted to conservation biologists Oswald Schmitz and Thomas Lovejoy for the consummate attention and advice they gave to the project in this critical period. I am deeply indebted to manuscript editor Susan Laity for contributions to the final product that were transformative. I thank freelance writer Alison D'Addieco for her technical assistance in manuscript preparation.

There are others during these years whose wise counsel resulted in material improvements in the manuscript: Greg Cooper, Harry Corwin, Laura Dice, Amy Eckhardt, Carolyn Flamm, Louise Gaffney-Gross, Amanda Gregg, Gene Gruver, Michael Hurwitz, Sumir Pandit, Leonard Plotnicov, Mike Rosenmeier, Daniel Ryan, James Simkins, Michael Skirpan, Mary Jo Wilson, Judy Zang, Mark Zang, and Nathan Zimmerman. Scores of students over the years read the book in my course, and numerous Honors College students have read and discussed it with me. I've mentioned by name those whom I recorded as most significant for their concrete impact upon the book. No doubt I have left some out, and to them I offer my apologies.

For all of the revisions and fine advice over the five years of its development, the manuscript has remained almost the same length as when first written, and it moves through the same topics in the same order. Still, no one could read this book then and now and not realize that it is supremely improved. Whatever the shortcomings in this effort, they are certainly mine. I am indebted to a veritable orchestra of supporters for making this project so much more than it could have been in clarity, organization, and precision. I can only hope that the result represents a worthy contribution to the reader's imagination.

To an Inquisitive Mind Open
to Honest Reflection, the Value of Every
Species Is Incalculable

I t may be only we ourselves, humans, who have the cognitive grasp to see other species for what they are. So it is ironic that of all living beings we would be the ones extinguishing other species from the earth. But that is what we are doing. Many experts believe that as a consequence of human activities countless species alive today will be gone by the end of this century. Estimates vary widely, both about the number of species on earth and about the number lost in recent decades, yet even with this uncertainty, there is reason to believe that the present pace of extinctions may rival that of the fastest mass extinctions in the earth's history.[1] Exacerbating the problem, populations of organisms are declining at even faster rates than species, jeopardizing the safe-guards of both redundancy (a large population has a better chance of survival) and the diversity within species that allows them to adapt to changing environments.[2]

What do we mean here by *species?* There are a variety of plausible definitions, but regardless of which you choose, the fact remains that species are disappearing at a rapid pace. This exceptional situation does not rest on technical distinctions of nomenclature that are unfamiliar to ordinary citizens. The concept of "a species" is a part of everyday vocabulary, and that is the meaning used in this book. When I speak of species, I am referring to "kinds" of organisms grouped according to shared physical and genetic traits. At times I may focus on subspecies or races, but my main concern is not with fine distinctions of classification. What is important is simply that by *species* I mean the kinds into which biologists divide *living things* as the best account of the nature of existence, the best explanatory typology. We all know that there are ascending levels of organisms within this typology—species, genus, family, order, and so on—and we know that species are kinds at a primary level whose details closely correspond to those of the individual organisms that we encounter.

It is often said that the rapid loss of species today is a crisis. But let us play devil's advocate here. Could any of us explain to a skeptical person *precisely* what that crisis is? To be sure, other species contribute abundantly to our material well-being. But if that is our concern, then talk of a crisis may be premature. The species that are vanishing may not affect our material well-being at all. Our lives scarcely intersect with those of many other species on the earth, and although some level of biodiversity is critical for human survival, only a percentage of the species that we do encounter will ever serve our practical needs. Loss of particular species may not even matter to certain ecological systems that support humans. Further, we tend to protect the species that are beneficial to us. So perhaps the extinctions occurring today do not present a crisis for humanity after all?

What if these extinctions considered en masse did present such a crisis? Even then, would our concern really be for the species? Let us concede that we face severe risks to our material well-being and even to human survival if these trends continue, for clearly we do.[3] The biologists Paul and Anne Ehrlich grimly warn that this destruction of species is analogous to someone's popping rivets from an airplane wing because he or she felt certain that "the manufacturer made this plane much stronger than it needs to be."[4] These massive extinctions present a danger to the health of the earth's ecosystems as conservation biologists understand that idea. They are a threat to the "land community"—soils, waters, plants, and animals—described by the conservation pioneer Aldo Leopold.[5]

In that case, however, what we have is a crisis for our material well-being or for the earth's human carrying capacity or for the health of the land community, not a crisis because species as such are being lost. That would constitute a crisis only with regard to losing the species that were necessary for particular needs, whatever those species might be. How many of the endangered species are necessary? Which ones?[6] In effect, our "crisis" would be the losses of certain key species en masse, not the individual losses. A wide range of individual species extinctions might not matter at all.

On the other hand, some people maintain that the extinction of *any* species is a loss for humanity, regardless of material considerations. What might be the basis for this belief? If it does have a rationale, that rationale is apparently not common knowledge, for people all over the world are destroying species through their activities with little obvious understanding or concern. It almost seems absurd to promote every species as having a "value" for humanity when so few people bother to protect that value in practice.

Whatever the truth may be, this uncertainty about the value of saving species and the general failure of will when it comes to doing so are surely sealing the fate of many other inhabitants of the earth. If there is a crisis of extinction that we must address, the first step would seem to be reconciling our own priorities and motivations with regard to it. Beyond seeking legal or economic solutions, we must first and foremost resolve the crisis of who we want to be as humans.

In this book I seek to demonstrate the value of all species to the human condition, independent of their service to our material needs. Let me make this task very clear. Imagine a skeptic who claims that an appreciation of species in themselves is merely a matter of personal taste—like a fondness for particular car models or flavors of ice cream. If this skeptic were right, then such an appreciation would have no claim to special protection in the public interest. How should we respond to such a skeptic? That will be the first, most important issue I shall address in this book.

I shall also explore the broader landscape of values that compete with species preservation. Values do not reign in seclusion but contend with one another as options among choices. The value of biodiversity in any situation is always "biodiversity versus what?"—for example, biodiversity versus property rights, or biodiversity versus employment, or biodiversity versus other incentives that threaten it. However, it might also be preserving biodiversity versus postponing human hardships until they have become far worse. In this book I seek to sort out the fundamental features of this landscape of values. As we reflect upon the issues that are raised, my hope is that we shall find a greater understanding and sense of purpose for preserving them.

Let us precisely frame the initial issue. Many forms of life have value for us because they serve particular uses—for example, providing us with nourishment or making up the biodiversity necessary for human survival. Certain organisms in forest watersheds help to purify the waters in rivers, lakes, and reservoirs that yield our drinking supply. Particular species of tree provide wood for construction, paper for writing, and rubber for tires. Some plants have medicinal properties that help restore health. Pets give us companionship and protection. Flowers brighten our celebrations, and a variety of plant species enhance landscaping for urban scenery. Geese produce the down used in our pillows. A number of plants and animals supply the raw materials for much of our clothing. Wildlife promotes ecotourism, which accounts for half the global tourism market,[7] and provides us with recreation through hunting and fishing. All these uses supply the jobs and the income that drive human economies throughout the world.

We see many reasons to appreciate other forms of life and living landscapes for their instrumental values. These reach across every sector of civilization in a variety of ways—from food production, medicine, fuel supplies, construction, crafts, jobs, and international trade to aesthetic services, recreation, watershed protection, pollution control, climate control, genetics technologies, regulation of soil fertility, management of pests and diseases, and nature-based tourism. Clearly we should protect the biodiversity that is necessary to secure these practical benefits.

But what about the myriad species that have no practical use to us? Do they have some other value deserving our protection? Do forms of life have a noninstrumental value to us simply for what they are in themselves? This is a crucial question as we confront the rapid pace of extinctions today because so many species do not seem to have any uses that humans find serviceable. We might call such value an "intrinsic value" of species, but I prefer the term "inherent value" to dissociate my meaning from the idea of a value that belongs to the thing itself independent of human perspective.[8] For better or worse, we can only perceive the world ultimately from the most knowledgeable grasp available to our human perspective, regardless of what we attribute in the process to gods, fiats, or particular texts. Were a divinity to exist, it still is only through our human perspective that we interpret anything about that divinity. Only through human sensibilities can we apprehend any value for ourselves.

We can learn through these sensibilities that there is a transcendent value for us in species. That "inherent value" would not be a value that is beyond our human perspective, but a value that arises from something that all people should find notable in the nature of the thing that is valued, regardless of its practical uses. What sorts of qualities in other species might constitute an inherent value?

One suggestion is that aesthetic value might be such an inherent value in species for us. Aesthetics has been called the philosophical study of beauty and taste.[9] But what constitutes aesthetic value can be a question more of our attraction to appearances than of what we discover by studying something in depth. It is not clear how every living thing would necessarily have a claim to much rank in this realm of aesthetics.

For example, some geological formations have aesthetic value in the sense that they are "beautiful" to us, and certain geological formations, such as the Grand Canyon, may be much more beautiful to many people than numerous forms of life—such as the star-nosed mole with its face of pink tentacles or the turkey vulture with its featherless head of scarlet skin—could be imagined to be. Does this mean that those geological formations have more inherent value than these forms of life? Or might it mean instead that conventional aesthetics is not the proper measure of the inherent value of living things for us?

In this discussion, aesthetics is not the proper measure of value because aesthetics, as conventionally understood, is really about appeal to our sensory tastes and predilections of form, not about the value of living things in themselves, irrespective of these tastes and predilections.[10]

It is difficult to imagine what the value of a living thing might be apart from its instrumental values. For that matter, it may be difficult to imagine what the value of anything might be apart from its instrumental value. Yet unless we acknowledge that every living thing has such a value, our skeptic will prevail. Appreciation of countless species will turn out to be a matter of individual taste with little merit for the general public interest. What, then, could qualify as a value of something "for what it is in itself"?

Just when we might want to give up our search for such a value, one such value might come to mind for our consideration. This is a value that is within our reach all the time, seemingly unworthy of notice. When we do notice it, it captures our imagination and takes us over, yet it passes unnoticed or even disparaged in the blind rush of our everyday lives. It is, in fact, grievously misunderstood because of that neglect. For this is a value that makes us human, the wellspring and arbiter of human enlightenment. It could be the most compelling and profound reason that humans, of all living things, should care about the fate of life on earth. *Individual species are phenomena in this world of such intellectual moment—phenomena so interesting in their own right—that this alone gives them a value meriting human embrace.*

Intellectual interest can be sparked by anything in the universe, regardless of (or beyond) its utility: species, landscape features, technology. Consider a meteorite—for example, the Murchison meteorite that smashed into Australia in 1969. This special rock is still being investigated with much excitement because of its plethora of diverse organic compounds—compounds that must have rained down on the early earth and may have given rise to the earliest life. But this meteorite is not being investigated for any practical value it may have. That is not why people care about it. Meteorites are deeply interesting to us in themselves, and the Murchison meteorite is a good example of something in our world that is worth preserving purely for that reason.

Consider another example. An allochthonous mountain is a

mountain that originated somewhere other than where it is, such as Wyoming's Heart Mountain, whose upper portion is more than 350 million years old and yet rests mysteriously on material that is only about 55 million years old. How did the colossal upper portion of this mountain relocate virtually intact from its place of origin, perhaps twenty-five miles or more away? This has remained a burning question for more than a century, but only because of its scientific interest, not because the answer may have any practical implications.

Consider a third example, a gyroscope inside a submarine deep in the ocean. In this instance, the great intellectual interest of the gyroscope goes hand in hand with a practical human value. If the navigators know their starting position and have a speedometer, then with the gyroscope they have all they need to figure out their vessel's precise orientation, velocity, and planetary position at any time wherever they go. This startling power makes the gyroscope, indeed, interesting as a phenomenon of physics. In fact, the gyroscope's intellectual interest may even outweigh its stupendous economic value. How might we defend such a claim? Well, the economic value of a gyroscope cannot explain why it becomes, in the quest for principles of understanding, something wondrous to contemplate.

These three phenomena achieve such interest for us only when we learn about them, either about their unique particularities—those of the specific Murchison meteorite and the specific Heart Mountain—or about the theoretical characteristics that they share with others of a kind—those of gyroscopes. It seems that in each of these cases, we somehow intuitively and instantly make an assessment of the nature, range, and depth of the phenomenon's intellectual significance, and we are moved. In other words, there is a value for us here.

Champions for this deceptively artless value are long overdue, and we should ask ourselves, "Why would we be timid about upholding it?" Think carefully about this. It is a matter so basic and simple. An object of profound intellectual interest sparks a theoretical energy that awakens our curiosity and motivates its focus. That is no small feat. The propensity to astound us intellectually is beyond our control; it entails a power to grasp and be stirred by reality beyond us. Somehow we of all creatures have brains capable of such a response. Indeed, this faculty of ours to recognize and appreciate the intellectual value of special things is a deeply mysterious and striking phenomenon in itself. To what end would we brush this discerning faculty aside?

We might as soon brush human enlightenment aside. In all cases, phenomena become interesting to us relative to what we discover about them. This attraction for us, in every case, bears a fundamental relation to the inspiration we have to learn, to the stirring of our imagination to surprise, reflection, and a sense of wonder; and above all, as we shall see in the next chapter, to the stimulation of our imagination into honest curiosity about the world around us. Notice that "the world around us" includes the things that are familiar and ordinary, for sometimes we overlook the ordinary things. In effect, their ordinariness robs us of the ability to see their true qualities.

If we discover that all species are essentially interesting to us, it will be because of what we have learned about the living things that make them up. For the concept of species is one of living beings, divided into "kinds" that best explain what they are. Consequently, if species in general have a special value for us simply for what they are in themselves, this value must arise transparently from what it means to be "a living thing." So let us now reflect upon some of the varied dimensions of "what it means to be a living thing," and consider what makes "a living thing" something amazing.[11]

The "life" that all living things possess may be compared to a flame that passes from an individual to its progeny through reproduction. Any living thing is an instance in which a single flame of life remained aglow all down the line of its ancestors. Every forebear throughout the entire line survived to reproduce.

Consider the implications of this. The original ancestors of all life on earth arose as much as 3.8 billion years ago. Consequently, every plant and animal alive today is the forward point in a seamless continuum of life extending back in time throughout most of the history of the earth. It is difficult to conceive of an idea of this magnitude.[12]

This unbroken continuum of life carried forward through each living thing entails successes in natural selection that occurred at every moment of environmental impact upon every individual in a lineage extending back all of those billions of years. So every living species today signifies an accomplishment in survival that is virtually beyond intuitive comprehension.

The external form of each "living thing" bears much of the brunt of its impact with its environment. But making up each living thing are parts, such as hearts or chloroplasts, that emerged over time to create or support that external form. Those parts have their own parts, which also have parts, and so on through cellular, genetic, molecular, and further levels. The interrelated functioning of all these constituent parts making up the functioning whole is also integral to every form of life.

Further, these functional parts represent extraordinary success stories of their own. For example, the complex genes that are shared by humans and fruit flies have independently survived more than 570 million years since their common ancestor swam in the oceans.[13] Every functional part of a living thing signifies continuous survival in a flow of environments spanning an immense magnitude of years that collectively participated in its presence today.

An obvious and remarkable aspect of living individuals is their continuity of identity through a lifetime of change. An oak tree and its predecessor acorn are a continuous living thing. A moth and its predecessor caterpillar are a continuous living thing.

So are a dragonfly and its predecessor nymph, a frog and its predecessor tadpole, a leopard and its predecessor cub, a human adult and its predecessor infant. Every living thing in its lifetime moves through a seamless continuity or metamorphosis, all the while remaining the same living thing. This coherence of living individuals through time and process is so ubiquitous that we tend to take it for granted. Indeed, every living thing retains its identity even as the matter and cells composing it change continuously throughout its life.

Hereditary transmission is an essential part of being "a living thing." The cells of all living things contain microscopic genes bearing precise hereditary codes. These genetic codes carry the explicit design of the entire living individual, and they also carry latent, unexpressed design characteristics that may become expressed in descendants. Every form of life received this genetic endowment from its forebears and typically will contribute it to the process of reproduction that will yield its offspring. In the case of sexual reproduction, all the chemical and behavioral dynamics of mates finding mates and consummating unions come into play, and those dynamics also follow genetic directives.

If we consider only these chemical, biological, and behavioral processes of hereditary transmission, we see that what it means to be "a living thing" involves almost infinitely more than meets the eye. Whether you are gazing upon a plant, a bird, or a reptile, countless unseen genetic encryptions are there that determine, in interaction with the environment, both the physical characteristics and the behavior that you see. The details that we perceive of living things are always merely the effects of this great network of integrated functioning.

Every characteristic of a living thing is a story of ancient origins. This is a story of changes in form, generation by generation, minuscule degree by minuscule degree, in which the earth's environments have shaped successful adaptations through natural selection. In effect, a flow of events over billions of years helped build the forms of life we find today. The resulting adaptations of species to environments involve details of functional design at every level of structure that we are only beginning to understand. These adaptations are another component of what constitutes "a living thing."

Such adaptations over billions of years of natural selection entail ecological relationships between each form of life and the physical environment. These connections extend across the earth as a spatial dimension of living things that augments the temporal dimension of their evolution through time.[14] Ecological relationships are an integral part of every form of life, for the design of a living thing for adaptation cannot be understood without understanding these relationships.

Consider the pronghorn, an animal in the western grasslands of the United States that is commonly mistaken for an antelope. The pronghorn is one of the fastest animals on land in the world for the speeds that it can sustain over short as well as very long distances. This animal can sprint at sixty miles per hour and run at speeds of forty miles per hour or more over many miles. No predator sharing the North American continent can even approach such ability. But where did this exceptional adaptation for speed come from?

It turns out that long-vanished enemies still pursue the pronghorn with every bound, for they all evolved in conjunction with one another within ecologies of open plains over millions of years. "American cheetahs" disappeared about ten thousand years ago, but the legacy of these swift cats and those broad landscapes still resonates in the fleet-footed pronghorn.

What, then, is this phenomenon that we call "the pronghorn"? It is, in part, countless chases of cheetahs past. The fact that American cheetahs disappeared thousands of years ago is not pertinent to this observation, for the point would be the same if the American cheetahs were still around. Every detail of every living thing on earth resonates with a flow of past influences that helped create that detail over billions of years.[15]

There is more. No other evidence of life has been found in the universe that is as complex as some of the simplest life on earth. Beyond the earth, life itself seems to be exceedingly rare. Further, because every particular life form is the singular result of an unrepeatable series of events and complexities in evolution, life elsewhere in the universe will not duplicate life on earth. This makes each form of life on earth—every kind of bird, mammal, and plant—even more significant. Every species is something that is unique and unrepeatable in the universe.[16]

Most spectacular, we are "living things" ourselves. These dimensions all characterize our own being. Each of us carries forward an unbroken continuum of life stretching back billions of years into the earth's antiquity. In fact, we and other species of life on earth are all part of the same family tree. We all originated from the dawn of life on this planet those billions of years ago and share that extraordinary living kinship. In that sense, an interest in any living thing is an interest in the same dimensions that constitute our own nature.

We rely on detective work in paleontology to map out the past forms of life and the major challenges they faced. Now we also have genetic investigations to supplement this work. Techniques in DNA analysis permit us to trace for any living thing its degree of relationship to other forms of life and the timetable of its branching evolution. The combination of this work with paleontology is dramatically enlarging our understanding of the history of life on earth and our connections to other species.

Many forms of life engage in chemical, visual, or auditory signaling—be it the choruses of crickets and toads on a summer night, the head-bobbing displays of certain lizards and birds, or the pheromones released by ants on the prowl and moths in flight. This is another important dimension of being a living thing. In most cases this interaction is programmed by genes. But it also may be grounded in learning, as is the case so significantly among humans. Our learned foundation for communicating with other humans and some other forms of life is a fascinating fact about us, and about them.

Indeed, in our case this aptitude is so exceptional that it makes possible vast transmissions of cultural information from generation to generation through language and other vehicles of expression.[17] Representations of other forms of life figure pervasively in these human cultural communications ranging across literature, folklore, religion, art, science, and other constructs around the world and throughout history. In effect, humanity has shaped and been shaped biologically and culturally by other life on earth. Thus many species around the globe bear in their nature the imprint of humans and our ancestors, as our own species evolved along with them in shared ecosystems throughout the ages.

Let us now summarize these varied facets of what it means to be a "living thing" on our planet. Every living thing on earth today—every fish, beetle, mammal, and tree—is the forward point in a flame of life that has remained aglow throughout billions of years of impacts in the earth's history; a functional organization of parts serving parts at every level of detail that arose and survived in that endless tumultuous flow; a coherent identity that remains the same organism through a lifetime of changes of matter and form; an intricate coordination of capacities for hereditary transmission to future generations of life; the expression of count-less microscopic codes that program a changing physical form and behavioral characteristics throughout its lifetime; a reser-voir of adaptations to environments within a fabric of mutual adaptations among living things across the earth; in many cases, a communicator across reaches of space with other living things through complex synchronized capacities; and in every case, a phenomenon that evolved through a fathomless stream of events to become unique and unrepeatable in the universe.

With all this in mind, we return to our fundamental query: Do all other forms of life have inherent value for humanity?

There is one simple answer to this question: every living spe-cies represents a dynamic process of the earth that is infinitely astonishing. One can scarcely comprehend the depths of fascina-tion presented by each species. Indeed, living things are so as-tonishing by any reckoning, and in so many ways, that if this itself could point to a value for human intelligence, that value would surely be magnified exponentially in the case of species. And we all know that this astonishment does indeed point to a value of all species for us.

In fact, what is remarkable about this situation is how acces-sible the answer is to our question. This is not rocket science. Every living species represents for our awareness a profoundly captivating phenomenon that holds endless mysteries of relations to other life and to billions of years of evolution and our own origins. The only requirement for understanding and appreciating this value is an inquisitive mind open to honest reflection.

We might call this an intellectual value, but is there not emotion felt in this discovery? Biological diversity has been deemed the "living library" of the life sciences. Collectively, species today provide the principal repository of clues to the resilience of life across the face of the earth through its billions of years.[18] We are only now learning how to read these clues.

In effect, the biology of every living species represents an alternative pathway of success in surviving millennia of onslaughts. In losing just one species to extinction, we lose from this living library a life form that is separated from its nearest kin by thousands to millions of years.[19] Among humans of every age and for all time, are there not also emotional losses here of phenomena that are spellbinding in all their singularity?

Just as the pronghorn is partly a reflection of the presence of American cheetahs, every form of life represents the working of ages of the earth's career that have come and gone. Each feature of a living thing reflects a vast story about events in our planet's ancient history through untold realms of natural selection. Merely as a focus for our thoughtful attention, this renders every form of life on earth a constellation replete with interactive attributes rooted in billions of years of earth processes. No more compelling phenomenon in the universe for human contemplation, inquiry, or discovery is imaginable.

TWO

The Intellectual Value of
Species to Humans Stems from
Our Unique Character

W hat do we gain by recognizing the full intellectual value of species? The attraction we find in this discovery is not one of practical values. Indeed, the breadth and depth of the forces that make up living things characterize rats, mosquitoes, and poisonous plants no less than cows, timber, and corn. Our appreciation of these amazing dimensions is not predicated on a service that is performed for our practical needs. There is a value that we find in living things simply when we understand what they are in themselves, with all the relations to other life that their beings entail.

This special value is not an absolute value that inheres in

creation itself, beyond the scope of human beings.[1] How could such a good be known to us, and how could it be *ours?* Any value for us is a human value, and the inherent appreciation of living things in human intelligence belongs to us in that necessary sense. But this value also appears to be ours in a special concrete sense— namely, this appreciation may uniquely exist in our own species. Let us explore this idea.

Many living things have worth to us because of their instrumental roles in our sustenance, shelter, health, safety, or recreation. This is the case for every form of life. All living things find instrumental values in other living things. The list is endless. Trees provide anchors against the elements for mosses, bromeliads, and orchids. Microscopic rotifers give viruses a place to replicate. Flowering plants host insect pollinators. Burrows of gopher tortoises shelter rattlesnakes. Certain mites hitch rides on beetles to find food. Some plovers in Africa clean crocodile teeth. Ants look after aphids and milk them for honeydew. Wolves and ravens seem to collaborate in hunting. Milkweed immunizes monarch butterflies against predation. Microorganisms aid all mammals in digestion. Blue jays warn prey of predators. Young chimps and baboons tussle with each other to test their own behavior and learn each other's. The "worth" of these values for the respective living things in each of these cases is tantamount to the functional value of nourishment, shelter, survival, or play.

The merit of living things as objects of our inherent appreciation is a different matter, unrelated to practical needs or emotional or romantic views of the universe. Instead, this appreciation arises from our human capacity to learn what living things genuinely are unto themselves based on what we observe about them as we are prompted by curiosity. Let us examine for a moment this extraordinary force in our lives, for it is our window into an appreciation of species.

What we call curiosity is fundamental in human consciousness. And true curiosity always assumes an honesty in its exercise. This creates two conditions, curiosity and honesty, that are vital to our enlightenment about the world, and inherent in each is that it is logically bound to the other. How could we be intellectually honest without a degree of inquisitiveness to awaken our attention in the first place and to exercise and test our reasoning? Alternatively, how could we be curious about something yet not seek honest information? In "curiosity" here, we are essentially considering the motivation inherent in thought itself.[2]

How could life without this motivation of curiosity in the passing turmoil of experience, a desire to frame and answer questions and resolve puzzles, acquire an honest or even coherent perspective? What could one then be honest about, and how would one show it? On the other hand, how could disregard for the truth characterize a curious person? The philosopher Susan Haack has insightfully described how these concepts, with scores of others, map out the human ideal of "intellectual integrity." As she puts it, "Someone who is really inquiring into a question wants to discover the truth of that question, no matter what the truth may be."[3] It seems clear that curiosity and intellectual honesty are fundamental threads of our character that need to run together.

Our appreciation of other species in all their complexity is sparked by these allied capacities of curiosity and honesty. Curiosity in its human elaboration is the singular attribute that opens the door for us to intellectual values of every kind. These especially include values attendant on our sense of wonder and astonishment in discovering the world around us. Indeed, the very idea of astonishment or wonder without curiosity is inconceivable.

At the same time, these attributes of curiosity, wonder, and intellectual appreciation in their human refinement are a notable evolutionary development. We now realize that the human articulation of these attributes and their attendant values is intimately associated with the prodigious facility of our species for representational communication through language—that is, explicit communication with one another about our experiences. No other life form appears to possess our exceptional aptitude for representational communication and the associated inclination to reflect upon the world inquisitively and with wonder. The question can thus be raised: What would survival without a robust curiosity and sense of wonder imply for meaningful human existence?[4]

Intellectual integrity requires that we regularly challenge our assumptions, rather than follow our habits of thought wherever they lead. It is always easy to go along with our kind, much as other creatures do. Many animals are intensely preoccupied with the messages of their own species, and we are no exception. Human sounds and signals of all varieties overwhelm our daily experience and seize our attention with a kind of urgency. Often they interfere with the balance and accuracy of our judgment, and it is important to block them out from time to time for the sake of maintaining a healthy perspective.[5]

Imagine a person who is absorbed with the signals of civilization gazing into a clear and moonless night sky. The canopy of stars gives us no sense of self-affirmation, but it is real and makes palpable for us certain ultimate fundamentals. The sight of the stars can regularly remind us that we are in the grip of an infinite mystery on a planet spinning through space. Many priorities that are urged by civilization may not be consistent with such a soul-thundering realization, and awareness of it can make us uncomfortable. Even misleading priorities provide some security of direction for us, and it is not easy to turn away from that security. But which is more important—the comfort of moving in just any direction or seeking to live an honest life?

To retreat from such an encounter involves a kind of denial, a jettisoning of curiosity and honesty for the sake of some level of illusion. Respect for curiosity is partly a reverence for real life. It requires honesty in our reflections over blind allegiance to comfort, stability, or a kind of mystical romanticism.

Appreciating another living thing for itself requires this same discipline of honesty. Spiders and snakes are infinitely more fruitful objects of study than our fears of them allow, yet these fears saturate much of our culture. In fact, such fears can take on lives of their own, which we know as phobias, and they come in many versions. There are phobic reactions to spiders, snakes, heights, openness, crowds, blood, flying in airplanes, public speaking, and more. Phobias are excessive reactions arising from organic and psychological factors that are often largely extraneous to the objects of fear themselves.[6]

One cannot walk away from a phobia. It is not something to apologize for. We should respect the phobias of ourselves and others. But we must not use phobias to justify revulsion with a form of creation or to denounce a form of life. A mature, reflective person understands that intellect and emotion are not the same. An honest response to a phobic object would be, "My heart jumps out of my chest when I see that thing, so keep it away from me. But some day I will try to learn about it."

Much like phobias, anthropomorphic impressions continually distort our perception of other species. Consider the turtle, the worm, and the octopus. Anthropomorphism will probably favor the turtle, because it has two "arms" and two "legs." But this preference is not about something intrinsic to the turtle, it is about *us*. Avian vocalizations that we romanticize as "birdsongs" may be simply the efforts of birds to attract mates or defend their territories. "Adorable" bears actually like to bury the rotting carcasses of their kill and then dig them up and eat them, and many bears would like to do the same to us.

Such distortions in automatic responses to other life forms are universal in the animal world. They are inescapable. A cat does not react to a mouse as you do, nor does a cattle egret react to a tick as you do, a mongoose react to a cobra as you do, a porcupine react to a bear as you do, an orca react to a shark as you do, or a llama react to a dog as you do. The essential being of a mouse, a tick, a cobra, a bear, a shark, or a dog may have little to do with these instinctive reactions by cats, egrets, mongooses, porcupines, orcas, llamas, or humans. In each case, these are largely species-specific reactions disclosing genetic or cultural factors in the reacting creature.[7]

Our encounters with other living things should partly involve thinking about our reactions to them. Our spontaneous reactions tell us something about our own species organically and our own varied cultural learning. Often these responses mask blind spots in our self-conceptions, such as our inflated self-image, which enables us assume that our reactions to other living things are about *them,* and not about *us.*[8]

Honest curiosity about other forms of life could never erase or supplant our anthropomorphism. Despite the boundlessness of our scientific inquiries, we cannot escape the narrow prism of organic and cultural artifacts through which we view the world, with all of our mysterious and idiosyncratic perspectives. We are destined forever to find ourselves stepping into and out of our spontaneous human reactions, feeling an emotional phobia for spiders, for example, even as we learn that there is nothing to frighten us, or finding the calls of some birds pleasing or annoying even as we discover their true function.[9] This perpetual immersion in our human nature makes each of us a compelling object for our own curiosity whenever we reflect on other life forms.[10]

Here is another example of an unrealistic, phobic reaction. For its size, the hognose snake may be the most frightening creature in North America. If you stumble upon this reptile, it will flatten its neck like a cobra, inflate its body and hiss fiercely, while striking violently. Surely we are right to feel panic at this sight, you will say. Not so. The snake's behavior may be designed to evoke such a reaction, but humans need have no such panic. The hognose snake is completely harmless to them. It is not venomous, its strikes miss the mark, and its mouth is generally closed when it strikes. The snake will not bite you, even if you pick it up. Instead, it will jerk convulsively and then roll onto its back with its tongue hanging out, "playing dead," as we say. The hognose snake has merely evolved in such manner to survive encounters with the likes of us, and it is no more dead in this case than it was dangerous before.

We begin to learn the true nature of other living things only when we attend to our curiosity instead of our gut reactions. Curiosity alone opens the door to the appreciation of life on earth, including ours, for what it actually is. To remain simply frightened of snakes or simply enamored of birds to the exclusion of any effort at honest curiosity leaves us imprisoned in emotion and instinct, inhibiting our self-realization.

Curiosity and intellectual honesty are fundamental conditions of a vital mind. They create in each of us the potential for growth and renewal. Surely we all know this, but let us observe just how deeply these currents flow in us. These two attributes of character enabled us to develop as "persons" by acquiring the language through which we communicate with others. They facilitated for each of us the construction of our self-awareness from the moment that our consciousness began to emerge in infancy.[11]

We could never have learned the meaning of a single word without honest communication because meaning and truth are logically bound together. In a general sense, the "meaning" of a word is its contribution to the "truth" of assertions that contain it, a profound and momentous idea. Thus we infer a word's meaning from its role in true sentences. Those who communicated authentically with us in words during our first years of life thereby gave us our ability to communicate in words with others and our ability to think in a rich, reflective manner.[12] As for "curiosity," it animated our attention and motivated our thinking about what was said to us. In a very real sense, each of us has been honestly curious about the world ever since we began to acquire language.[13]

These observations lend momentous spiritual significance to what is at stake for us in losing the variety of life on earth, for here are losses that seem to touch the essential and distinguishing privileges of our humanness itself. In play from our earliest spoken words are attributes of human consciousness that yield an utterly breathtaking new chapter in the cosmic biography—the emerging power of a life form able to reflect with insight and wonder upon life itself in all its infinite dimensions. A bat will never know what an astonishing thing it is. Nor will a crow, or an ant, or a sloth. We stand alone among all living things on earth in this gift of appreciating every species for its uniqueness. Further, we see that we share within our very DNA these same relationships to earth events, and even a living kinship with all the earth's species in an extraordinary genealogy. It is this path of human enlightenment for its own sake that opens for all generations the realms of value inherent in each species.

This perception does not mean that we must save every other living species at all costs, for that would be impossible. Some species cannot be saved. Indeed, some species will become extinct before we know that they exist or are endangered. What it does mean is this: in the contest among our values, the value of other forms of life for the human imagination and intellect should be accorded the weight that we place on associated qualities of character that we hold in reverence—our curiosity and our honesty, for example, and our sense of respect and awe before creation.

Think again of the brilliant canopy of stars that is visible on a moonless night. It is seemingly *completely useless* for our practical needs. But where is the person who would stand by indifferently if we lost forever our view of the stars?

It is well to lay out these points with care and accuracy. There is too much at stake for any confusion here. This is not a crisis concerning the practical values inherent in all living species, but a failure to recognize the values of perception that distinguish our human imagination from all others. We stand alone as a form of life that is able to reflect upon the natural world with wonder and to realize new orders of gratification in pursuit of that awareness. It is a form of life that resists submission to stereotyping and pragmatism to seek out the awesome dimensions that natural things genuinely represent. But we must now awaken to the fact that our civilization is destroying species throughout the world on a devastating scale. Indifference to this destruction of the life of our planet is patently unconscionable when our distinguishing qualities of intellect and imagination are considered. To be indifferent is to deny one of the most salient features of humankind.

The rapid pace of species loss today is not a threat to "nature." That is not where the crisis lies. Life rebounded over the course of millions of years following each of the five major extinction events in the earth's history. It is likely to rebound again over millions of years from whatever damage humans might inflict during our own horizons of time.

Neither does this crisis consist of a threat to values found outside the human perspective, for we cannot understand those values. Rather, the crisis concerns the fate of values of other forms of life to fundamentally human sensibilities of consciousness. In a very real sense, species loss is a distinctively human version of self-destruction.

Some discussions of "sustainability" dangerously miss this point. It is said that we must save the jungles because there may be undiscovered cures for cancer there. It is said that loss of biodiversity may reach a tipping point that threatens human survival. These are resonant arguments, but offered alone they present sustainability for health and survival as if the only value of other living things is their instrumental value for us.

Who honestly doubts that we could lose wild animals in Africa, North America, and every other continent without jeopardizing human survival? Who honestly doubts that we could lose mammals in the oceans without jeopardizing human survival? Who honestly doubts that we could lose the orchid family, the world's largest family of flowering plants, without jeopardizing human survival?

Human survival as a goal represents an exceedingly narrow ambition that is hardly commensurate with the full scope of human nature. Missed in the narrow calculus of human survival is the value of living things to curious and honest people for what they are in themselves. A wholesale destruction of other species in the family of life that has evolved over billions of years would represent an incalculable loss of the earth's heritage for mindful generations of the future.

That would be a monumental tragedy for the human condition in any scenario. Sustainability as a goal must not just be sustainability for mundane human survival. It must also be sustainability of the earth's living heritage for a flourishing human spirit.

THREE

The Fate of Life on Earth Hinges on Property Values

The value of the earth's living heritage to our humanity is incalculable. No honest and reflective person could deny this. There may be no avoiding our destruction of some forms of life. But if we are to find a way to at least diminish that destruction, it will most likely emerge only from a recognition of the limitations of reasoning and the self-constructed boundaries that place us in jeopardy of losing so much.

In this chapter I turn to examples of these limitations and boundaries as they appear in competing values in land. Threats to species often arise in the context of attitudes toward land that define our view of the world and our interactions with the earth and with one another. These attitudes vary from people to people, although their connection to our shared human or animal nature may also be strong. They may have emerged centuries or even millennia ago, long before the devastating impacts of today's crowded earth would have been anticipated. Like it or not, here we are.

The institution of "private property" sets the context for a great deal of angry politics over the fate of environmental assets in much of the world. This is inevitable, for we share "private property" everywhere with other living things and with ecological resources. The reason values in species create controversies for property law is not that people can own species as private property and dispose of them as chattel. A person cannot own naturally occurring species, as distinct from individual organisms. Instead, the reason values in species create controversies for property law is that people can own the land where species are distributed and, by that ownership of realty, can dispose of the land and of most of its organisms.[1] So let us examine the institution of land ownership and associated values in the world's most influential developed nation, the United States. Understanding exactly how U.S. property law works and the values that arise in conflict with it is essential to our stage for the chapters to come.

American readers should find that the values discussed in this chapter ring true, for this is their culture. But a wider, international readership should notice that this examination also resonates for them in familiar ways. The legal culture of the United States is gaining ground throughout the world as market economies spread. We need an example to see this clearly in action, and an apt one can be found in the nation of Mongolia. We will see later how this new nation vividly demonstrates the impulses to privatize resources that are in play throughout the world.

But even for parts of the world where privatization is not an issue, a discussion of private and public tensions over land use in the United States can be instructive. Such tensions will probably eventually arise in these regions, and they are likely to resemble those in the United States and thus to implicate the fate of other species. Indeed, understanding the fundamentals of property law and values in the United States should help us see how sentiments that drive land use in this country reflect general characteristics of human reasoning and motivation that imperil species in every corner of the earth.[2]

In the United States, so-called property rights advocates carry much of the fight against federal and state laws protecting wetlands and endangered species. This is because it is with rights in private property that those laws compete. Or do they? It does seem logical, this idea—that one can protect the public interest in wetlands and endangered species only by competing with rights in private property.

But although this view may be convenient politics in some quarters, it misrepresents the law. Whether in legal history or in common sense, the "public interest" always ranks supreme. Consequently, any time a government defines and protects a system of private property, this too must be subordinated to the determination of the public interest. In that sense, "property rights" by definition can never exceed the limits that the public sets to safeguard the public interest. Laws genuinely protecting the public interest in wetlands and endangered species could not compete with any property rights that anyone possessed.

In the concept of private property, a government creates certain individual powers for citizens to use against the intrusions of others, including itself. This is no trivial assignment. For some in America, it is a core guarantee of freedom expressly embraced for the public interest by the nation's founding fathers. Nonetheless, its potential threat to the earth's living heritage and the horizons of future generations is undeniably profound.

If all property on earth, including the oceans, were privatized, the fate of the earth's living heritage would hinge entirely on private property rights. This intimate association of species survival with property rights presents challenges for human nature that have had unintended tragic consequences.

The actors in this longstanding duel over the reach of property rights in America today are the U.S. Constitution and common law. Property rights advocates make a banner of the "takings clause" of the Fifth Amendment of the Constitution, which prohibits government from "taking" private property for public use without compensating the owner. However, the takings clause provides only one side of the picture, and it is patently absurd without the boundaries of the other side: the "police power" provided under common law. At the root of both are our intuitive conceptions about what is fair. Driving these conceptions of fairness are our values.

To help us examine these relationships, let us first look at how the relevant elements of American property law work. The principal source of property law in the United States is common law, which is to say, law arising in court opinions based on principles in prior opinions. This series of common law principles, each resting upon earlier enunciated principles in judicial holdings, extends back in time through the law of American courts, and before them through English courts, and sometimes from those back to Roman law and the origins of dispute resolution itself. How does this ancient legacy of common law affect the fate of other species today whose lives may compete with human purposes?

One common law principle in the United States that is explicitly traceable at least back to Roman civil law is the idea of police power, which asserts the supremacy of the public interest over individual rights in every realm and provides governments with the authority to enforce that supremacy. Enforcing the supremacy of the public interest is the purpose of governments. Police power is therefore as old and legitimate as government itself.

Rights in ownership of property are not exempt from police power. Police power implies that private property "rights" end where the public interest mandates. It thereby defines the boundaries of private property as a cultural construct. No citizen can obtain rights in ownership of property that intrude on the public interest or, in legal jargon, create "nuisances." Consequently, no limitation in ownership of property under police power to secure the public interest denies a citizen, or "takes" from a citizen, any rights that he or she possesses.

Here is a famous statement of police power made in 1851 by the chief justice of the Massachusetts Supreme Judicial Court:

> We think it a well settled principle, growing out of the nature of well ordered civil society, that every holder of property, however absolute and unqualified may be his title, holds it under the implied liability that his use of it may be so regulated, that it shall not be injurious to the equal enjoyment of others having an equal right to the enjoyment of their property, nor injurious to the rights of the community. All property in this commonwealth . . . is derived directly or indirectly from the government, and held subject to those general regulations, which are necessary to the common good and general welfare. Rights of property, like all other social and conventional rights, are subject to such reasonable limitations in their enjoyment,

as shall prevent them from being injurious, and to such reasonable restraints and regulations established by law, as the legislature, under the governing and controlling power vested in them by the constitution, may think necessary and expedient.[3]

There is a second principle under U.S. common law distinct from police power that provides the government with further authority over private property. This is eminent domain—the power of governments to forcibly purchase individuals' rights in private property in order to serve the public interest. In the case of eminent domain, governments make individuals relinquish their property rights for a payment of value in order to accomplish public objectives. In the case of police power, governments compel individuals to terminate property uses because they are "nuisances" and individuals have no right to those uses.

Eminent domain stops individuals' activities on their private land through forced payment in the name of the public interest. Police power stops their activities without payment in the name of the public interest. In the former, there are no nuisances, and consequently there are property rights and payments for them. In the latter, there are nuisances, and consequently there are no rights to be bought and no payments. The two doctrines are independent and mutually exclusive.

The judiciary must determine in both cases whether governments are acting in the public interest. Courts are usually asked to decide which kind of authority the public interest at issue merits—police power or eminent domain. Thus, through common law there arises the definition of the "public interest" itself, as courts set the scope of police power versus eminent domain and affirm or deny governments' assertion of police power over individual rights in particular situations.

Different accounts of the public interest draw different boundaries on the institution of private property. Underlying them all is a contest of values, a struggle for the future of our descendants. The fate of other life on earth hangs in the balance.

The institution of private property as shaped through common law in the United States consists of a "bundle of rights" to a piece of land. For example, a "life estate" is a bundle of rights to a piece of land strictly for the duration of one's life, at the termination of which another legal estate outside one's control and vested in another individual takes over. "Easements" and "leaseholds" are bundles of rights to land within estates belonging to others.

The usual interest that is acquired when land is bought is called a "fee simple" interest. In the case of fee simple ownership, the bundle of rights acquired is the maximum bundle possible under the law and includes the following general sorts of rights:

- right to occupy
- right to exclude others

- right to use
- right to subdivide
- right to build upon
- right to sell
- right to bequeath
- right to lease or rent

Because of the doctrine of police power, nowhere in this bundle of rights is there this right: "right to create a nuisance." Because of the doctrine of eminent domain, nowhere is there this right: "right to stop government from forcibly buying this land for a public objective."

The requirement of a fair price when the government exercises eminent domain is underscored in the U.S. Constitution itself. This is the "takings clause" of the Fifth Amendment that property rights advocates invoke: "nor shall private property be taken for public use, without just compensation."

Here in the takings clause, the Constitution does not require compensation whenever governments restrict private property uses in the public interest. That would eliminate police power and with it the very essence of government. Under police power in common law, the government does not have to pay property owners in order to prevent them from creating nuisances on their land. The common law doctrine of police power remains in full force alongside the common law doctrine of eminent domain to which the Constitution's takings clause applies. The Constitution does not replace common law in this case; it supplements it.

The legal question that is raised when government denies a use of private property, such as a use that endangers a species, is this: Is it a police power action to prevent a nuisance, or is it a takings clause/eminent domain action that requires just compensation?

More important, a decisive ethical and political question accompanies that legal question: When the government terminates certain land uses by an owner of private property, should the owner get nothing by way of compensation (police power) or should the owner receive fair compensation (takings clause/ eminent domain)? Courts throughout history have answered this question by differentiating, in case after case, situations of nuisances that permit police power from situations of eminent domain in which police power does not apply.

The least controversial opinions hold that nuisances encompass dangers to public health or safety. This reflects a public consensus about values. We are likely to say that when a landowner uses his or her private property in a way that endangers public health and safety and the government stops that use, the landowner should not receive compensation because landowners do not have the right to use their property in such a way. No rights have been taken.

What about landowners who use their private property in a

way that endangers another species? What results when this is prohibited? Courts in almost all cases regard these new prohibitions to be exercises of police power as well, and this result is very unpopular in many quarters. It is frequently challenged.

A property owner can challenge government restrictions on the use of private land and will probably retain an attorney to do this. The attorney will argue that the restriction is an unconstitutional taking of private property and look for as many cases as possible in which controlling or reputable courts have held that similar restrictions are takings.

The government will counter that the restriction is a legitimate exercise of police power and seek to find as many cases as possible in which the courts have held that similar restricted uses are nuisances. The court must decide which doctrine applies: either there is a Fifth Amendment taking, in which case the restriction is unconstitutional unless just compensation is paid, or there is a legitimate exercise of police power, in which case the property owner receives no compensation. The result will be fair compensation or no compensation. It is all or nothing.

Courts faced relatively little public controversy in identifying nuisances until well into the twentieth century. According to one crude formula, police power is a matter of "preventing a harm," while the exercise of eminent domain is a matter of "conferring a benefit." Through much of history this formula seemed to make sense. Exercise of police power prevents land uses that pollute groundwater with carcinogens or air with noxious fumes, or that put children in danger, such as uncovered wells, dilapidated and fragile buildings, or refrigerators with unbolted doors. These restrictions are comfortably regarded as preventing harms rather than conferring benefits. Exercise of eminent domain would include building highways or power lines that cut across private property, building dams that flood property, and so on. These actions confer benefits rather than prevent harms.

Today, however, this proves to be a weak and elusive distinction to bear the weight of absolute consequences for payment of landowners. Simply put, preventing a harm does confer the benefit of preventing that harm, and conferring a benefit does prevent the harm of denying that benefit. So this formula has fallen by the wayside in modern analysis, and no clear replacement has been found.[4] Still, courts have faced little public controversy when distinguishing certain police power situations from takings clause/ eminent domain situations because of the relatively settled public consensus about the particular values connected with those situations. With the advent of values affecting the fate of other species, there is no longer that settled consensus.

The rise of environmental regulations that protect not only health and safety but also species and habitats has led to a much larger realm of restriction on private land uses than ever before. In this new realm, courts generally expand police power to accommodate the new environmental legislation. But there is no clearly established public consensus to support doing so, as in health and safety cases. As a result, the opposition heats up intensely, to the great detriment of other forms of life that inhabit the same property. The underlying values tell the story.

Consider the following propositions:

1. We must not pay people to refrain from activities that create hazards to public health and safety on their private land.

2. It would be nice if we could pay people to refrain from activities that create hazards to public health and safety on their private land.

3. We must pay people to refrain from activities that create hazards to public health and safety on their private land.

Among these three propositions, the public would probably choose the first. Creating hazards to public health and safety on one's private land is not regarded as a right. Police power is the correct option to prevent it.

Now consider these propositions:

1. We must not pay people to permit public works like power lines on their private land.

2. It would be nice if we could pay people to permit public works like power lines on their private land.

3. We must pay people to permit public works like power lines on their private land.

Here, the public would most likely choose the third option. Preventing public intrusions such as power lines across one's private land is regarded as a right entailing some protection. Eminent domain is the correct option to achieve those public works.

Finally, consider the following propositions:

1. We must not pay people to refrain from activities on their private land that contribute to the demise of another species.

2. It would be nice if we could pay people to refrain from activities on their private land that contribute to the demise of another species.

3. We must pay people to refrain from activities on their private land that contribute to the demise of another species.

Among these three, the public would most likely choose the second option. There is no consensus about whether engaging in activities on one's private land that contribute to the demise of another species should be a right. Deciding whether police power or eminent domain is the best way to prevent activities on private land that jeopardize other species means deciding whether to compensate land owners.[5] Protecting other species does not stand up against private property rights nearly as well as protecting health and safety does.[6]

What is surprising about these preferences is not that protecting human health and safety ranks above species survival as a public priority. People will lay down their lives in law enforcement and military service to protect the health and safety of other human beings of all races and nationalities. Far fewer will do so to protect other species. That is an interesting fact about us, but it is not surprising. What is surprising is that the institution of private property has any political weight in the battle against initiatives to protect endangered species.

Notice what this regard for private property means. Sentiments about property rights are rather capricious. An investment in any asset carries risks, including the risk that actions by the government will injure the investment. Consider the effects upon stock value of selective sales taxes or government procurement decisions. There is no feeling that governments should compensate stockholders for their losses in these cases. People concerned about such risks can invest in assets that they think would be more secure than these. Why should their attitude be different toward real estate risks involved in protecting species? That they are is rather surprising.

An investment in land to grow crops for export carries the

risk that the government will embargo the exports. An investment in land for markets supported by tariffs carries the risk that the government will remove the tariffs. An investment in land alongside a prominent highway or a military base carries the risk that the government will redirect highway traffic or close the base. Technically, government actions like these do not restrict what one can do with one's land, but as a practical matter they certainly do. Why should the assumption of risk become so objectionable when the government *directly* bars certain land activities, even to protect species, if the effects of indirect measures are often the same or worse?

Most startling, the lifespan of every human being is trifling when compared with the larger aspirations of humanity. The length of time that any person owns a parcel of property is even shorter than a lifetime. Indeed, as with any other investment, the period of ownership can be as short as the individual chooses it to be. Still, many people are inclined to give individuals the right to reduce the living heritage of the earth for all future generations no matter how briefly they own a piece of property—even if it's only for a week. That is not just surprising, it is absolutely astounding.

Nevertheless, this is the case. In the United States, the institution of private property as received through common law has a political power to stand in the way of protecting endangered species and similar earth assets that it does not have against our other human interests.[7] Property rights retain this power against protecting endangered species regardless of how briefly the property is owned in a lifetime by the person claiming those rights.

Let us look at the restrictions placed on logging under the Endangered Species Act to protect northern spotted owls in old-growth forests of the Pacific Northwest or the red-cockaded woodpecker in old-growth forests of the southeastern United States. Are such restrictions a legitimate exercise of police power to prevent nuisances, as U.S. courts generally would hold, or are they unconstitutional takings of property?

Even a timber landowner who is an "environmentalist" might be inclined to reject the court's stand on police power here and respond, "Each member of the public at large benefits from the continued existence of the spotted owl and red-cockaded woodpecker through my not logging my property, but the law makes a few landowners like me bear the entire cost of that public benefit. That is not fair. Everyone should share the cost. I should be compensated for my loss from the public coffers."

But a factory owner could retort, "Each member of the public at large benefits by continued good health from my terminating industry on my property that pollutes the air, but the law makes a few landowners like me bear the entire cost of that benefit. That is not fair. Everyone should share the cost. I should be compensated for my loss from the public coffers."

It is hard to imagine anyone embracing the latter argument, in sharp distinction to their response to the former one. In that former case, the owl and woodpecker themselves become lightning rods for public anger against the government as can be seen in slogans on automobile bumper stickers.[8] Comparable attitudes attacking public health to spite the government would be inconceivable. In the case of public health, an alternative argument to "fairness" wins the public consensus: "The landowner wants to reap his personal land profits at the expense of the public interest by jeopardizing public health. That is not fair. No one has the right to use his or her property that way."

Yet why do we not also say, "The landowner wants to reap his personal land profits at the expense of the public interest by jeopardizing the survival of other species. That is not fair. No one has the right to use his or her property that way"?

Again, what is surprising is not that protecting human health and safety ranks above the survival of other species as a public priority. It is that rights related to the short-term, voluntary choice to own private property have such political weight in the fight to protect the earth's living heritage for all future generations.

It is even more remarkable that the owl, the woodpecker, and like assets have become targets of anger over limitations put on property rights to save them. Attempts to protect these species seem to put them at greater risk. With something like contempt for other forms of life, the public is ready to sacrifice them in the battle with government over the limitation of property rights, of all things. Imagine what might happen if people angry over police power restrictions on their use of property were to make public health and safety a target.

This observation is not to take sides in that battle; the fact of the battle itself is worrisome enough, let alone the question of whether it can be won. It is to recognize what appear to be self-destructive impulses in humanity as the steward of the earth's heritage for the good of posterity. These impulses animate the very nature of our practical reasoning.

Humans Are Poised to Destroy
the Resources of a World
of Bountiful Interest

R ational self-interest is a defining condition of human nature. Rationality and self-interest are inseparable in the realm of practical reasoning—reasoning about what one should do in any given situation. Consider a simple test.

Imagine that you place your hand on a hot stove and pull it away, and I ask you why you pulled your hand away. You'll respond that the stove was "hot." If I then ask you why the fact that the stove was hot is a reason to pull your hand away, you'll respond that hot stoves "hurt" you when you touch them. If I now ask you why the fact that the stove hurt you when you touch it is a reason to pull your hand away, you will respond that this last question is ridiculous, that I must not understand what "a reason to pull your hand away" means. The fact that the stove hurts you is reason in itself to pull your hand away without further explanation for anyone who knows what hurting oneself means. There is simply no more that can be said in response.

The question of why things that hurt *me* should concern *me* makes no sense. The question of why things that hurt *you* should concern *me* does make sense. Your pain is not my pain, and my pain alone is the logical starting point of my practical reasoning.[1]

This fact about the logical starting point of practical reasoning is the same for each of us. From this starting point, concern for others must be constructed and explained as part of our practical reasoning. Practical reasoning is ultimately about maximizing "personal benefit."[2]

From this first principle of rational self-interest, the economist Garrett Hardin in a famous essay identified a fundamental problem of human nature that he called the "tragedy of the commons," a concept that is so important that it deserves to be repeated here.[3] Suppose there is a "commons"—a realm of benefits of any kind whatsoever—that is *open to everyone,* and that each person has *unrestricted access* to take these benefits. The realm of benefits has a "carrying capacity"—that is, the benefits remaining in the realm after others are taken out can be sustained to a lesser and lesser degree as more and more are taken, until all the benefits are exhausted or destroyed.

We would like to think that any rational person who had unrestricted opportunities to take benefits from that realm would, as a matter of rational conduct, refrain from taking benefits that are necessary to ensure a future supply. This is not the case, however. In fact, as long as the realm of benefits remains completely unrestricted, our human nature, guided by rational self-interest, will lead us to overrun the realm's carrying capacity and exhaust the benefits. Indeed, it could be considered irrational for any individual left to his or her own devices to limit the removal of benefits from an unrestricted commons, even though the person knows that doing so is necessary to sustain those benefits.

That refraining from using up the benefits in a commons could be both necessary for human survival and humanly irrational is remarkable. But if it can be true with regard to benefits tied to our survival, how much worse is the case for benefits tied to our values in earth's living heritage independent of human use.

Hardin's example is that of a hypothetical herdsman confronting a field of grass that can support a limited number of sheep. The herdsman is deciding whether to add the first sheep to the field. Doing so will give him milk, meat, and wool. Doing so will also reduce the amount of grass in the pasture available for sheep. Now the herdsman's gain from the first sheep will be greater than the loss he will suffer from that sheep's reduction of grass in the pasture. So the herdsman should introduce the first sheep. What about another sheep? Same calculation, same result.

And so on, sheep after sheep, until the carrying capacity of the field is reached.

The wise herdsman would stop at the point of carrying capacity, but only if the pasture belonged to him for his exclusive use. But this is a commons, and everyone is allowed to add a sheep. The herdsman in the hypothetical example will thus reason that if he does not add another sheep, some other herdsman will do so, and the grass will be lost without his gaining any benefit from the sheep, and so consequently it is pointless for him to restrain himself. Indeed, once he realizes that the pasture is both completely unrestricted and finite, if there is no limit to the sheep that herdsmen can use he should add as many sheep as he can as quickly as he can so as to beat other herdsmen to the chase. Even at the carrying capacity, it would be irrational for the herdsman not to add another sheep, and quickly, unless he has an assurance that other herdsmen will stop adding sheep at that point as well. Otherwise, his not adding a sheep will not save anything, but it will deny him some final benefits before disaster strikes.

That assurance about what other herdsmen will do would need to be a reliable understanding—based on traditions, ethics, or law, for example—among the herdsmen as to how the field of grass is to be used. And that understanding would constitute a public restriction on removal of the benefits, a restriction that we assumed as part of the hypothetical does not exist.

This inexorable logic of the tragedy of the commons can

operate not just in pastures but in every realm of the earth's resources. Individual decisions that deplete resources include having one more child, adding one more unit of pollution into the air, sending one more boat out to fish an ocean, adding one more home to a scenic landscape, cutting off a switchback on a hiking trail, or taking one more life from a species that is endangered. Without some form of dependably observed public restrictions— that is, without some level of assurance that we shall all coordinate our choices to save the benefits—human nature (individuals behaving rationally with regard to a public commons) will lead us to destroy all the earth's resources that human beings need for survival or desire as part of a world of bountiful interest.[4]

Whenever the results that we seek from our actions require people to coordinate their behavior, insecurity about whether others will do their part can interfere with our plans. The greater that insecurity, the less reason we or anyone else will have to behave in the desired way. Reliable public restrictions in the form of social norms and sanctions or government-imposed laws can remove that insecurity. In the case of the complex human challenges facing species today, the effective choice among these would be government-imposed laws.

Perhaps we could find a different solution. What if, instead of imposing government restrictions on our commons, we eliminated the commons altogether by privatizing all the earth's assets?

If our herdsman owned the field of grass himself, his rational self-interest would force him to consider the health of the range as a critically valuable asset that he fully controls. It would lead him to manage the range in a way that was consistent with its carrying capacity.[5]

Such a solution would work for the herdsman and his field of grass, but it would not work for species protection. Concern about a field of grass that is the herdsman's source of sustenance is not the same as concern about a species that is an object of intellectual appreciation. Unlike a field of grass owned by a herdsman, virtually no species is situated on a singly owned tract of land. Further, in the case of many species the individual members move around from tract to tract.

In that sense, the land is not the commons, the species is, so privatizing the land does not eliminate the commons. Since no single person owns a species and uniquely controls its fate, but rather many people control its fate, each person is subject to some uncertainty about whether others will protect the species and thus to uncertainty about the value of their own efforts to protect it. It would seem to follow that once a species is at risk from human practices, and thus the "commons" is already being exploited at the expense of the species, only enforced public restrictions can eliminate that uncertainty and facilitate the protective behavior sought from rationally self-interested individuals.

If our concern is with protecting grasslands, on the other

hand, we know that virtually all grasslands, if separately owned, will be protected by rational herdsmen because their survival will depend on their doing so. Survival is the fundamental goal of every form of life on earth. Survival, for herdsmen, is a matter of maintaining sustainable grasslands. So we might think that there is no need for enforced public restrictions beyond mere ownership by herdsmen to protect grasslands.

But a herdsman protecting his own grassland might behave in a way that injured the health and safety of other herdsmen, such as using toxic insecticides to kill pests, and such behavior would require enforced public restrictions on land use. Still, we do not need restrictions simply to keep pastures from being overgrazed by herdsmen who own them and have no other livelihood. Knowledge of the land's carrying capacity plus an instinct for survival will prevent that.

Yet protecting species for their inherent value to us is not a question of our survival. Consequently, privatizing land does not protect species for their inherent value unless we can find some other rationale. Indeed, securing the resources that we believe we need in order to survive could endanger all assets that compete with this, including species. We could make this point about the herdsman and his pasture. He needs grass and ease of livestock movement and therefore may decide to target competitors, such as prairie dogs and badgers. For prairie dogs also eat grass, and the burrows of both animals uproot pastures and allegedly endanger livestock. Wholesale and deliberate destruction of prairie-dog towns by ranchers in America's western prairies pushed to the brink of extinction the black-footed ferret, an animal that lives entirely in prairie-dog towns. This small mammal once numbered as many as eight hundred thousand individuals across the Great Plains, but it was thought for many years to have been annihilated in the massive campaign against prairie dogs. In 1981 a single population was discovered that yielded just eighteen ferrets for captive breeding.[6]

If we imagine grasslands as being owned by persons whose survival depends on rentals for the scenic view, there will be a different outcome. The owner may want to replace the grass with other landscaping, possibly including voracious exotics that renters will find attractive. In addition, the owner will not want to mar the landscape with too many rental units, so there is a carrying capacity of benefits here to be respected.[7] In any event, survival rules the day.

There is plainly no incentive comparable to survival to ensure that people will protect species as assets in themselves. Indeed, species are bound to be destroyed even if they merely conflict with "inconvenience" survival values, such as influencing a property owner against his will to change the precise nature or location of his livelihood. Only enforced public restrictions can protect the values of species and other ecological assets from conflict with values like these—that is, the various orders of survival values.

There is another argument in favor of government protection of species. A new and awesome force has entered this calculus. With the rise of civilization, a far-reaching institution has developed that is universally compelling for humans, even eclipsing survival values in various ways—*the institution of money in a global marketplace.* Global networks of commerce are capturing ever-widening realms of commodities today, and the consequences are genuinely revolutionary for human choices among values. Without money, it seems, no person can even satisfy basic needs, but with money a person can now obtain virtually any value in the world.[8] So powerful is money today as a means of access to innumerable values that some people will risk survival itself for a chance at a fortune.

Absent enforced public restrictions on the use of private property, any rationally self-interested owner of a property will seek to protect and enhance its monetary value against all opposing values, including the value of other forms of life. This is a major new threat among the challenges we face. No effort to protect other species can ignore the competing value of money given the nature of human practical reasoning.

Property Ownership and the Desire for Money Work Against the Interests of Species

However superior its inherent merit may be, the value of another living thing may stand at great disadvantage against other values, especially the value of money. This can follow from the very nature of our practical reasoning and motivations.

A sack of money is not a perishable value of single dimension, like a sack of grain. Instead, money represents an enduring power to acquire innumerable values of all levels and kinds—from survival and sustenance to pleasures of every variety, quality of life, and social status. One might even say that accepting a monetary

bribe in exchange for destroying a species would be a simple matter for many people. Indeed, it happens frequently.[1]

It follows naturally from this that converting the benefits of the commons to money greatly magnifies the tragedy of the commons. If the sheep in our pasture are only to feed and clothe our family, then there is a limit to the number of the sheep we can use. But if the sheep can make money for us, and we have access to a global market for them, then the number of sheep that we can use increases exponentially, and so does the pressure on the pasture. Indeed, our incentive to protect the carrying capacity may even diminish in some cases relative to the size of the short-term profits to be made and the option of taking them elsewhere. In effect, as money gains more purchasing power in the global market economy, the conquest of things from the earth that make money expands indefinitely with market demand.

At the same time, the horizon of our values remains foreshortened. While monetary values encompass near-term values and long-term values alike, our incentives are weighted toward the near term. So that is where monetary incentives largely gravitate. In deciding what to do, as a matter of practical reasoning, we have a tendency to maximize our personal benefits, but with a bias for the short term rather than the long term. We are more stirred to action by outcomes expected immediately than by those projected, even with certainty, a few years away. By nature, we are

little concerned with consequences that are deferred very far into our future.[2]

Consequences beyond our lifetime are much less compelling for us than those within our lifetime. Consequences for our children and other descendents whom we know are much more compelling for us than those for potential descendents whom we do not know, such as great-great-great-grandchildren. Outcomes projected to arise more than a century from now scarcely find a place in our thoughts. Indeed, a person who devoted substantial attention to outcomes in the far distant future would be an object of derision.

All of this is not good news for species, because threats to their survival usually present themselves as playing out in a fairly distant future. If that future is presented as a century from now, the threat will not receive much priority among humans. If that future is a quarter century from now, its value must compete with values whose outcomes will occur sooner, such as all the many values we can achieve from having money.

Further, threats to the survival of other living things are usually realized degree by degree, species by species, rather than for a wholesale mass of species all at once. The immediate allure of money with all its purchasing power easily defeats the immediate loss of species that may be felt or threatened at any given time. In the case of species loss, the devastating cumulative impact is delayed until some vague future point in time, and that danger is much less compelling to us than more near-term, well-defined values.

Yet another aspect of human nature that gets in the way of protecting species is that we are more concerned with effects that take place close to us than with those happening farther away. Farmers who destroy cutworms that eat their crops in Kansas will probably not be moved by the effect this has upon grizzly bears in Wyoming. But grizzlies feed on the cutworm moths that have flown from Kansas to talus slopes in western mountains, and if the worms are destroyed, then so are the moths—and this threatens the grizzlies.[3] In general, very few species that we harm through our actions suffer the impacts in our vicinity.

All these problems play further into the logic of the tragedy of the commons that looms behind our choices. Remember that this logic arises when we try to make rational decisions about using an open realm of benefits in the absence of security that others will make their decisions in conjunction with us—Why should I avoid walking on the grass if I have no assurance that others will avoid doing so as well? Now consider any potential decision of yours that may threaten the survival of one or more species. The more uncertain the full extent of that threat, the more delayed into the future, the farther away the consequences, the less secure you should be that a decision to protect species will sway other people. This is especially true when the alternative is a value as definite, versatile, and timeless as money. And the less secure you are, the less rational it is for you to forgo those appealing and immediate monetary values yourself. The fact that we all presume that everyone is reasoning in the same way only compounds the problem.

In other words, as a general matter, the long-term prospects for humanity are disadvantaged by a pervasive human bias favoring the near-term gratification of each individual over the welfare of the human species. The quest of each individual for money is driven to serve the individual, not humanity. In turn, the institution of private property serves the quest for money. That bond of property and money dominates land-use outcomes today as never before.

Ownership of private property and the desire for money go together. Any nation seeking to maximize economic productivity for its standing in the world and the benefits that money brings to its citizens will find private ownership of land an essential.

An illustration would be helpful, and today a timely example of the way a newly introduced market economy is driving the privatization of property is the recently formed state of Mongolia. A traditional pastoral economy once characterized this ancient land. Now Mongolia is adopting a diversified market economy in a global system. Accordingly, many Mongolians are seeing prospects of greater personal wealth and a sea change in their relationships to land.

Let us examine why this is the case, and consider how other species throughout the world might fare if private ownership replaced common ownership to serve a market economy. We have seen that a completely unrestricted commons invites disaster for all the land's resources through the tragedy of the commons. How does privatization of property affect the land's resources in the global market economy?

Peoples of the North Asian steppe, where Mongolia is situated, have maintained a nomadic culture based upon animal husbandry for millennia. Their essentials of food, shelter, clothing, and trade have derived from pastoral animals—sheep, goats, yaks, camels, cows, and horses. Humans first domesticated horses in this region more than 5,500 years ago, and this was a profound development in human history. Domesticated horses opened the way for stable settlements, the building of cities, trade across vast areas, new levels of warfare, and geographic conquest.[4]

Pastoralism in Mongolia traditionally involved free-range grazing through vast and separate seasonal pastures under ancient nomadic traditions. The very dry and cold climate in Mongolia limits the capacity of a single range to support stock year round, so herding families must be on the move. Even the traditional Mongolian house, which is found throughout the North Asian steppe, is well suited to nomadic life, for it is collapsible and movable. (This house is sometimes called a yurt, but Mongolians prefer the term *ger.*) Its circular walls are constructed of several sections of crisscrossed latticework covered with felt. The entire structure can be assembled or collapsed within thirty minutes and fits atop a single draft animal.

The Western concept of private property is alien and antagonistic to this nomadic population. Pastoralism on the Asian steppe involves movement in traditional patterns over the land to rotate animals between seasonal pastures, often widely separated. In the United States, ownership of property is a means of protecting a private realm from intrusion by others, including the state. Freedom from state intrusion is a revered "liberty." But nomadic cultures conceptualize liberty as a freedom from barriers to movement. Nomadic liberty is not posed in opposition to the state but, quite the contrary, is afforded *by* the state's monopoly on land. Private ownership would destroy that liberty. This is exactly the reverse of Western thinking.[5]

Such openness to movement across the land is "the foundation of nomadic liberty," according to the commentator Mashbat Sarlagtay. His vivid elucidation of this idea provides a deep and palpable contrast with the values of land ownership that are disseminated across America, Europe, and much of the rest of the world today:

> Unlimited nomadic activity means that there can be no private ownership of land. Land in a nomadic society is like the air or the ocean, it is impossible to divide and

possess. It is not even public property, but simply a lim-
itless expanse where we live and move. Nomads want
to travel everywhere and across everything, without any
limit. Can you imagine their thoughts if a stranger ap-
peared before them, saying "This piece of land is mine"
and prohibiting them to go across it? To own a little piece
of landmass of the universe, saying "It is mine," sounds to
them like "this cubic meter of air is mine, so, you cannot
breath it!" It is impossible to imagine.[6]

Americans might pause in wistful contemplation of a celestial
liberty that is so obvious and yet so alien to their own frame of
reference.[7]

Soviet influence prevailed in Mongolia with the creation of
the People's Republic of Mongolia in 1924. This lasted for about
seventy years, when popular demonstrations demanded political
freedom and human rights. Mongolia established a multiparty
system in 1990 and held its first democratic elections. The republic
was renamed the state of Mongolia in 1991, and the nation opened
its first stock exchange and adopted a new democratic constitution in 1992. This constitution provides the mandate and the
groundwork for a privatized economy.[8]

Virtually all nations compete for eminence today in a global
market economy, and this economy is rooted in privatization. The
transformation of nations like Mongolia into global market com-

petitors is fueled by the immense appeal of the goods money can buy, from foodstuff and beautiful things to cars, televisions, laptops, and cameras. These are immensely attractive values, and the global market broadcasts them by satellite everywhere. These signals reach wind-powered televisions and personal computers in the world's most remote homesteads. They reach gers deep in the Mongolian desert.

Fundamental to this extraordinary situation is the concept of private property. In the first place, the purchasing power that provides these astounding goods and drives this captivating economy depends on money in the hands of buyers. The economic value of real property is by far the largest potential source of capital in the developing world.[9]

When an individual "owns" land, that ownership is protected through registration in records enforced by the government. Ownership of property implies ownership of the monetary value of the property based on market demand. Because the law protects land ownership, banks will take the risk of providing interest-bearing loans to landowners based on the value of their land and secured by mortgages. In this way, land privatization instantly provides otherwise impoverished citizens with the means (mortgage money) to create and amplify economic activity.

In addition, legally enforced private ownership of land provides incentives to focus purchasing power on land improvements, for the landowner will own those improvements and reap the benefits of the enhanced land value. Land improvements mean the property has higher sales value, rental value, and other productive values if the improvements are installations for better agriculture, mining, or other land-based industries. Without legally enforced land ownership, those using the land have no rational incentive to devote labor and capital to improvements, which the state or others might take away from them.

Legally enforced land ownership encourages ambitious and savvy business minds to compete for productive land innovations. Among competitors for a piece of land, the winners will more likely be those who know how to increase the monetary value of the land still further, rather than those who are less aggressive in maximizing productivity. That will optimize productivity in the nation at large, thereby raising both its global prestige and its revenue from taxation.[10] All these benefits require intense land development for the single purpose of generating more money.

The transition from values founded on nomadic liberty to Western land values founded on individualist liberty amounts to a paradigm shift in the relationship of Mongolians to their land. Mongolia is introducing property rights to a people who have never owned a piece of land privately before. The monumental nature of this paradigm change does not impede the transition to privatization—even a swift transition—provided that economic choices on a personal level are sufficiently compelling. It is a matter of human nature vis-à-vis the value of money that those choices are sufficiently compelling if the land can yield profits.

That seems to be the case in and around certain urban areas in Mongolia, where privatization is moving swiftly and enthusiastically in opposition to the traditional pastoral economy. Profits can be earned in favored locations by building apartments, motels, service facilities, and shopping centers. At the margin, some land will not yield income whatever the effort. Where this is the case, banks will not lend funds, and privatization offers no gain. As a general rule, privatization accelerates competition to convert land to profits wherever that can be done.[11]

Mongolia decided to exempt pastures from full privatization in a partial accommodation to its long traditions,[12] but this benefit is being offset by long-term pasture leases. In the outcome, the profit incentives actually work against protecting the range, just as one might expect. The herders tend to crowd herds onto their exclusively held sites, especially winter campsites and sites near water, leading to localized overgrazing. In short, with only a limited-term lease to a parcel that is too small for traditions of wide stock rotation to protect the range, a herder has less mobility and less reason to protect the pasture instead of extracting maximum profits for the finite lease term.[13]

The best examples of the way privatization in Mongolia is threatening social and environmental values may be the untoward effects of the mining leases distributed by the government. Mining leases are private interests in land that have economic consequences similar to those of absolute ownership. In Mongolia, as elsewhere, unregulated investment opportunities favor the short-term profits of investors over long-term quality of life. Mongolia has leased a substantial portion of its state-owned land to foster mining of its rich mineral deposits. These include the world's largest copper mine, with an exploration area of more than 34,000 square miles. Mining requires copious amounts of fresh water, which is diverted from communities and livestock. The ensuing water depletion and polluting effluents jeopardize public health and destroy local drinking water reserves and ecosystem support. Underlying all these effects is simply a new paradigm of private property, mining leases, introduced to a formerly nomadic culture. Far from protecting national resources from a tragedy of the commons, privatization wildly intensifies the short-term exploitation of land for monetary profit.[14]

Transition to private land values naturally serves the immense appeal of money, which empowers an individual to acquire all the popular values of the global market, short-term and long-term. For better or worse, this transition is likely to bring in its train the same conflicts between public goals and private goals that afflict Western law. As commentators report of Ulaanbaatar, Mongolia's capital city, "Now—in UB city—it is difficult to possess or use a piece of land for commercial purposes or for [a] summer cottage, because most of the valuable land is already occupied. For example, playgrounds for children, green areas, sports areas for the schools, and common lands do not exist anymore and they have been replaced by constructions such as bars, night clubs, restaurants, hotels and apartments."[15]

Once more and more people claim "rights to private property," however ephemeral their tenure on that land may be in the landscape of human destiny, those claimed rights are bound to collide with evolving public values. In the United States, we see this conflict in the tension between police power and the guarantee against the government's taking property without just compensation. Now we find the same tension formalized in Mongolia's constitution.[16] The idea that land ownership is a mere investment like any other investment, with unpredictable risks of loss, or that land owned now will pass through successive generations long after the short lifetime and personal choice of today's owner—observations like these can be lost when exalted "private property rights" dominate the rhetoric.

There may be a tendency for a "right to private property" to become exalted artificially in this manner, unless there is a conscientious design to keep that from happening. A freestanding "right to private property" may be a deeply persuasive idea for human nature in the context of the market economy, just as a freestanding "right to free movement" may be a deeply persuasive idea for human nature in the context of traditional nomadic cultures.

As Mongolia and other nations make the transformation to Western land values, they are creating new legal schemes virtually from scratch. These will define formidable new conceptions of property rights, social incentives, and civic stewardship in the collective consciousness of their future generations. These nations can never turn back from those consequences. At the same time, they can look to the time-tested example of the United States for a model of what needs improvement.

The global economy today propels a tremendous complex of accelerating market forces driven by the appeal of money to the earth's expanding human populations. This readily overpowers a cooler and more intelligent civic resolve. *Other species of life live on "private property" as well as humans.* But it is hard to see how the inherent value of the earth's living heritage to our humanity gains any effective voice in the destiny of private property against the juggernaut of money as value for more and more people.

As critical as is controlling human population growth, this challenge requires even more of us. In the first place, it requires precise limitations on property ownership. Here Mongolia has the advantage over the United States, for Mongolia's constitution expressly restricts property rights to protect sweeping environmental values.[17] In the second place, it requires a firm and evenhanded enforcement of laws that govern property allocation and use.[18] But perhaps most important, managing this challenge requires a consensus, renewed with each new generation, about long-term human values on this earth.[19]

Absent that, the idea of property rights will surely assume a life of its own, challenging legal restrictions under a commanding market mindset that places accumulation of money in competition with all other values. Such a state of affairs will jeopardize the survival of other species throughout the world day by day, and will easily overwhelm the intelligent national stewardship that ultimately defines and protects a worthy civilization.

SIX

Free Market Environmentalism
Places Profits Above the
Public Interest

I t may be tempting to conclude that the only way to rescue the earth's living heritage from competition with money is to coopt money and the free market into its service. Proposals to do just that under the rubric of "free market environmentalism" have gained a following among varied constituencies in recent years. These constituencies include individuals at opposite poles of the values spectrum, which may not augur well for the integrity of markets as arbiters of the public interest. As the economist Thomas Power has explained, at issue is the world of difference between having markets *determine* the values we maintain, and having markets *serve* the values we maintain.[1]

Consider the Clean Air Act passed by the U.S. Congress in 1990.[2] This act is designed in part to reduce sulfur-dioxide emissions into the air from coal-fired electric generators. The quantity of sulfur dioxide emitted from a smokestack is measurable. The goal is to decrease the annual nationwide emissions of sulfur dioxide from utility units to no more than a certain ceiling by a particular year. This means reducing the annual emissions by specified millions of tons during that time.

The Clean Air Act defines an "allowance" as an emission of one ton of sulfur dioxide per year. It provides for an automatic distribution to generators nationwide of a fixed number of allowances, equaling the amount needed to achieve the reductions; the allowances vary according to each generator's characteristics. A generator is permitted to emit only its assigned allowances of sulfur dioxide. The act also provides for the U.S. government to withhold a percentage of allowances from distribution to make available a reserve of allowances to distribute at market prices in a public auction.

It follows that under the Clean Air Act the annual sulfur-dioxide emissions by the target year should total no more than the designated ceiling even if generators buy and use all the additional allowances from the reserve. That result would meet America's goal. The act does not dictate the means by which generators must meet their emissions limit, and generators can transfer their allotted allowances from one plant to another, thus exploiting a surplus of allowances in a clean plant to fill a deficit in a dirtier plant. Companies appreciate this flexibility, and that favors their compliance. Further, every generator has an incentive to be cleaner than its permitted allotment of allowances because a cleaner generator can sell its unused allowances on the open market. That is a big improvement over former U.S. laws, which merely set standards to be met, with no incentives to exceed them.

So here we have a market mechanism serving the goal of cleaner air with improved incentives to achieve compliance. That is a good thing.[3] On the other hand, although the market *serves* the goal of reducing sulfur-dioxide emissions to a designated ceiling per year, the market does not *determine* that goal. That would be absurd. We do not let markets determine permissible pollution.

We do not allow markets to decide everything. Monetary incentives provide the purchasing power for so much personal gratification that they easily become the dominant motivation of behavior and overpower all nonmonetary values unless we make sure that this does not happen.

There are many nonmonetary values.[4] We do not let markets determine permissible pollution. We do not let markets entice murderers to publish memoirs. We do not let markets offer the right to betray. We do not willingly let markets provide the sexual services of children or trade in slavery. We pay public officials specified salaries rather than let market payments influence their decisions. We do not let markets determine our spiritual pursuits.

We do not let markets rule what we teach our children. We do not let markets determine ethics.

As a conceptual matter, letting markets determine ethics would not result in the achievement of "ethics" by any reckoning. Ethical evaluation of character or conduct does not permit exceptions based on a person's physical characteristics, social status, or material means. Being tall, strong, or socially powerful is not an excuse for being dishonest. Being *rich* is not an excuse for being dishonest, either.

Ethical principles are like a compact that each of us has a reason to embrace largely because we are confident that others in our community will embrace the compact as well. Being in the compact is its own reward, for we all benefit from the communal bond. Otherwise we would be fools to respect and embrace the bond ourselves.

Our dilemma in motivating ethical behavior is much like that of the tragedy of the commons, for here as well we need the security of knowing that other people will comply with the ethical compact in order to rationalize doing so ourselves. Enforcement of the compact accomplishes that. If someone deals dishonestly with us, we usually find out, and when we do, we'll have little reason to treat them with any ethical consideration (and they know this), and we will take action to enforce the consequences of our disapproval. Since everyone is behaving in this way, that keeps us all in line.

Permitting exemptions to the ethical compact based on a person's height, eye color, strength, political power, or wealth would weaken its rational authority for us. It would compromise the fundamental respect that we accord to the security of equal status for all under the compact that justifies its collective embrace. Emotionally, the response would be indignation if someone suggested that our honor could be bought. So we do not permit the market to determine ethics.

Instead, so powerful is money as a value for us that we must protect our ethical choices from the marketplace. The lure of money easily overrides our ethical judgment. When this happens, and we permit money to buy a lapse in ethics, we regard it as a kind of corruption. Thus, determining the "public interest" involves deciding what values we must protect from corruption by markets.

It is safe for money and markets to determine how our civilization manages matters of personal taste, such as choices among flavors or preferences in general merchandise. Whether a person chooses one thing or its competitor in this realm does not matter to the public interest. If more money can be made in the market from compact discs than from vinyl records and we let the one replace the other in production, there is no problem for the public interest. We have not let money corrupt our values. We have not sold out.

Similarly, if more money can be made in the market from wide neckties than from narrow neckties, from chocolate than from carob, or from digital cameras than from film cameras and we let the one replace the other in production, there is no problem for the public interest. We have not let money corrupt our values. We have not sold out.

On the other hand, if more money can be made in the market from having casinos available at school recess rather than play-grounds and we let the one replace the other, there *is* a problem for the public interest. We have let money corrupt our values. We have sold out.

So the question is, Which goals should we let markets deter-mine for us, as in the case of flavors and merchandise, and which goals should markets serve if they can but not determine for us, as in the case of pollution and public decency standards?

What we count as the corruption of our values depends on how we answer these questions. Consequently, so does the integ-rity of our social order itself. Indeed, how we protect and motivate our paramount values in a market economy is probably the most important issue confronting the public interest since money was invented. This is certainly our profound challenge in protecting other species of life.

A paper company that owns vast acreage in the United States

has replaced existing forests with pine forests for its products. Company executives realized that people might like to fish and hunt on this property and that selling permits for those purposes could generate income. Indeed, the market for these permits turned out to be so substantial that the company curtailed logging within buffer zones around streams and other areas to enrich the hunting and fishing.[5]

The outcome has been greater biodiversity in general within those zones, and this has surely benefited ecosystem services to forms of life well beyond the company property. All of this is a favorable result for environmentalism. But is this a case of a market determining goals or of a market serving goals embraced independent of market considerations?

It is most likely purely the former. The company's profit stream determines what it supports. If the demand for fishing and hunting were not cost effective, the company would cease protecting the buffer zones. As far as the company is concerned, if no one will pay enough to protect those areas from logging, that protection will not exist, for there is no other rationale to embrace it beyond fostering goodwill.

One might respond, "Look, the company is protecting those ecosystem services, so what does it matter whether this goal is financially based or is reasoned from the public interest?"

If the market supports a goal, does it matter whether the public interest makes the goal a priority as well if the two happen to be coincident? This question suggests that we might escape having to agree about the definition of the public interest in these cases, but this is an illusion. The question itself places the market in the driver's seat, for it supplants the alternative and more appropriate question: If the public interest supports a goal, does it matter whether the market favors the goal as well?

A passive reliance on markets to determine goals for our civilization in the absence of direction from the public interest removes the public interest from the table. It leaves only the profit stream and all that it might yield in the serendipity of circumstances. It removes all oversight of market effects. In the case of endangered species, it implies that it does not matter whether there is a public consensus about the value to humans of biodiversity, ecosystem services, and other forms of life.

The above examples illustrate two versions of "free market environmentalism." One version asserts that we should carefully design market mechanisms to motivate better compliance with environmental standards that we determine independent of markets. The Clean Air Act clearly shows that market mechanisms can help achieve ceilings on air pollution that we set independent of markets. That is a good thing for environmentalism.

The other version asserts that we should let markets determine what our environmental standards are because otherwise those standards do not matter. This is *not* a good thing for environmentalism. Indeed, this version cannot be called environmentalism under any plausible meaning of the term.[6] It would liken our goals for endangered species to our goals for merchandise and flavors, which are goals that we turn over entirely to markets because they do not matter. The answer to whether other species of life have any *inherent* value to humanity would simply be no.

A number of arguments supporting both versions of free market environmentalism draw compelling contrasts with the alternative of typical governmental management. It is said that even if the government determines the environmental values to be maintained, using traditional governmental management in achieving those values is counterproductive because it entails lethargic bureaucracy, inadequate accountability for efficiency and excellence, minimal local stewardship, absence of incentives to exceed standards, and the public's resentment of punitive, "command and control" authority. Both versions agree on these criticisms of government, and even environmentalists must admit that governmental management of natural resources can seem weak in these respects when compared with private-sector efficiency and incentives.[7]

However, other arguments that are presented to support free market environmentalism would have markets alone determine the environmental values we maintain. It is said that governmental management of public lands demands no separate fees for our varied recreational alternatives, and thus enables "no basis for ascertaining the true value of these services or the worth of the project or activity."[8] But notice that we do not let fees measure the "true value" of alternative educational standards or spiritual pursuits. It is said that governmental management is undemocratic insofar as it inappropriately assumes and subsidizes certain values, insulating them from a fairer marketplace forum where competing values could emerge.[9] But we do not let market competition determine our pollution standards or our slavery standards. Is that "undemocratic"?

A perspective of markets serving goals slides all too easily into one of markets determining goals when we introduce markets into decision making about environmental values. Consider this comment:

> Yellowstone experienced many growing pains between 1872 and 1892. Most of these problems stemmed from the difficulties of shaping the landscape into a marketable product. Many Americans, in fact many people around the world, embrace national parks as the best remaining examples of wilderness preserved, but that was never anyone's plan for Yellowstone or the other national parks in the early days. Indeed, the idea of wilderness was about to undergo significant revision and itself become a commodity.[10]

But while it may be true that the idea of wilderness has developed into a value for us during the past century, that does not make this idea a simple "commodity." This particular term has a cynical ring, and sometimes for good reason, as when it disregards important distinctions. "Commodities" are not worth much in our larger scheme of values.

We do not speak of educational objectives as "commodities" even though their values are associated with profits in the marketplace. Similarly, the money raised in fundraising against drunk driving or earned by taxi drivers from drunken revelers does not make our value of driving sober and the criminal statutes associated with drunk driving "commodities." National parks remain creatures of legislation, protected against the unbridled marketplace, notwithstanding their very effective marketing.

None of these values is a mere commodity. In each case, we have removed and protected the respective values from the unbridled marketplace. Educational and spiritual opportunities exist as the options they are in our society because of their treatment under our tax codes. Wilderness areas exist as options for us because of the Wilderness Act, signed into law in 1964. We do not treat these values as we do commodities like flavors or merchandise, for that would be the end of them. The idea of wilderness could not survive as a commodity in any free market. We may let markets serve the goals of these values, but in so doing we must be vigilant that the market stops there.

Why is this vigilance necessary? Why should we believe that a shift from markets serving our goals to markets determining our goals is a genuine risk for the public interest? There are several grounds for believing this. First and foremost, money is a tremendously seductive motivator that easily overwhelms other values, for the reasons I have enumerated. Each of us understands the power of money from a personal standpoint. And as it can always seem harmless to try to make just a little more money, there is a slippery slope here.

Second, the more each of us thinks that others will follow the money, the less reason there is for us not to do so as well if the alternatives require universal cooperation. Insecurity about the social fabric is not just a problem for the tragedy of the commons. Confidence in others is essential to ethical behavior in general. There is no reason to tell the truth if no one else is telling the truth. In the same way, choosing to follow the public interest becomes less rational for each of us the less secure we are that others will do so instead of making money at the public's expense.

There is a third important reason for worrying about the special influence of markets in our decision making. The mere fact that market motivation is permitted to operate in a realm of choices tends to disparage the use of better judgment.

Suppose that you paid people whenever they were honest with you. The implication would be that the payment was a reward for their honesty. They do not have to accept that payment.

But if they do accept it, this would imply that they also accepted that implication. A kind of agreement has now been reached with no exchange of words. If you then stop paying them, you are removing their new, implied reason to be honest with you. In effect, your earlier payments legitimize their no longer being honest with you unless you come to a new understanding with them. The mere fact that you paid them has disparaged their use of better judgment about why they should be honest. It was probably not a good idea to pay them whenever they were honest.

The moral is this: If you pay someone whenever they are honest with you, you had better make sure they know that the payment is not the reason to be honest. Similarly, if there is a better reason for an environmental value than the money it could generate in the marketplace, we had better emphasize that reason clearly, persuasively, and relentlessly if we want to protect the value once the market enters the picture.[11]

The idea of wilderness is especially at risk from an expanding market mentality because wilderness represents a value for which free market environmentalism will always fall far short. People who oppose America's Wilderness Act know that without it, the idea of preserving wilderness would give way to the much more lucrative goal of land development for logging, mining, and recreation. They are absolutely right. No one could challenge that. The Wilderness Act defines wilderness as "an area where the earth and its community of life are untrammeled by man, where man himself is a visitor who does not remain." The act generally bans from wilderness areas motorized equipment and mechanical transportation.[12] In effect, the Wilderness Act implicitly keeps much of the marketplace out of wilderness areas.

It follows that the idea of "wilderness" could not survive in any genuinely free market. Neither could other species of life as objects of our inherent appreciation. An unbridled marketplace could not be a true friend to these values. In that sense, the idea of wilderness and our inherent appreciation of other species of life are limiting cases of values that prove a point: free market environmentalism is a contradiction in terms. The public interest could not be protected in any marketplace that is genuinely and completely free. To believe otherwise and to place an unbridled market in control of our more considered values is to put the cart before the horse.

It is confusing and even contradictory when we find market values and public interest values operating together, for their perspectives are not related. But especially when these values operate together, it is important to remember that a failure to distinguish clearly between them places the public interest at risk, not the market.

Income streams that distort the values of public agencies with broad missions are prime examples. Thomas Power reports that in Montana and Wyoming fish and wildlife agencies that are funded by fishing and hunting licenses resisted reintroduction of wolves into the Greater Yellowstone ecosystem because they feared the loss of game and the associated income from hunting licenses. The Montana Department of Fish, Wildlife and Parks opposed listing the grizzly bear under the U.S. Endangered Species Act because that would mean an end to hunting licenses for grizzlies.[13] In effect, values competing with values that are fed by markets struggle while values aligned with markets flourish.

Game hunting offers an instructive case study. People enjoy hunting for many reasons, and this creates a market for property interests in game for shooting, interests that are measured by the payment of fees in the form of hunting licenses as well as by the huge commercial profits from hunting products. A hunting license amounts to a private ownership right in the hunted animal. Hunters do not want to hunt game to extinction in their own tragedies of commons, so they endorse legal restrictions on the numbers of animals they may kill each year. Violating these restrictions by killing animals without a license means criminal punishment for the violator if apprehended. But it also means condemnation by hunters themselves as conduct that violates the framework of property interests in which the hunters have personal stakes. Poaching especially angers those who outfit the hunters, for these outfitters earn their livelihood by means of hunters. In effect, killing game without a license amounts to theft of property from law-abiding hunters and their outfitters, and it occasions the wrath of both groups for this reason alone, if not others. Certainly the market forces involved in hunting are very effective in protecting these game animals.

On the other hand, the government also provokes the wrath of some hunters and outfitters when it compromises these property interests for reasons of environmental protection. Laws like the Endangered Species Act prohibit, for nonmonetary reasons, hunting certain animals and thereby challenge the property interests of hunters. An example might be the prohibitions on hunting wolves in certain locales. Indeed, these prohibitions have actually denied hunters a property interest in an animal that kills game itself—a double blow. But absent a property value in wolves for hunting, this leaves only the strong arm of the law to protect the wolves, a law at odds with the desires of those who would rather hunt them. In this situation, some hunters and outfitters may view the taking of wolves in violation of the law as harmless since it does not injure any property interest of theirs. Some of them may even admire the poaching of wolves as a legitimate signal of defiance against the government and the alternative values that the law subsidizes. Those values are not measurable by the prices of licenses, the fees of guides, the sales of products, or any other earnings.

Environmental values that have no significant market value are vulnerable to destruction by competing market-based values unless the government enforces laws to preserve them. But the government faces attack if an adequate public consensus does not endorse those environmental values as being in the public interest. Public anger at logging restrictions to protect northern spotted owls and red-cockaded woodpeckers is a case in point. Legal protections of these endangered species and of "wilderness" and other environmental values will not prevail against market competition in the long run unless the rationale for preserving these values for the public interest amounts to a precise and accurately articulated argument that we continue to find compelling and regularly enunciate as a check on market pressure.

We are back to the issue of whether the status of the earth's living heritage for the public interest is sufficient to withstand the power of money. As human efforts to maximize profits threaten the remaining land on earth, it is folly to think that markets alone can aid the survival of a substantial number of other forms of life.[14] Who would buy a license to hunt a salamander or a beetle?

Federal and state initiatives to resurrect the Everglades ecosystem amid Florida's explosive development dramatize this contest over the public interest. Are the Everglades a "God-forsaken swamp" that we should reclaim from mosquitoes, rattlesnakes, and alligators, as once popularly viewed, or are these wetlands "one of the unique regions of the earth, remote, never wholly known . . . unique also in the simplicity, the diversity, the related harmony of the forms of life they enclose"?[15] If the latter is the case, can we put a price on such a value? Is there a point at which so many profits will be at stake that government action protecting the Everglades would no longer be acceptable? Or do the Everglades represent a value for the public interest that is *beyond measure?* How do we decide?

The institution of money has become a defining cultural force that is still relatively young in the evolving chronicle of humanity. How we protect and motivate our paramount values in a market economy is probably the most important challenge for the public interest since the advent of money. The fate of other species rests virtually entirely on our decision.

SEVEN

Species Have No Direct Claim for Consideration in an Ethical Community

We do not let markets determine ethics. Associating ethical judgments with monetary rewards would obscure and even contradict the authority of ethics. It would create an expectation of a monetary reward for ethical behavior and then a sense of a right to that reward. This would weaken the authority of ethics and raise the absurd specter of an ethics of the highest bidder. Our reward for being ethical is usually the return favor of ethical behavior directed toward us in our ethical community. To suggest by a monetary reward that this may not be enough would call into question the foundation of the ethical compact.

It is not ethical to conduct activities on one's private land that jeopardize public health and safety. Consequently, the idea of paying someone to refrain from such activities is abhorrent. Paying someone to refrain from jeopardizing public health and safety would weaken the authority of the ethical principle that using one's private land in that way is wrong.

The idea of paying someone to refrain from causing the demise of other species would similarly be abhorrent if causing the demise of other species were thought to be comparably unethical. In that event, the doctrine of police power would protect other species with little controversy.

Is there any way to infer from our ethics the inherent value of other species?

Ethical sensibilities do extend to other forms of life within a certain realm, but that realm is not helpful for the fate of species. The idea of paying someone to refrain from hurting animals, for example, is abhorrent, because hurting animals seems wrong to a decisive majority of the public. With few exceptions, our ethics does not give us the right to hurt animals. Thus, many governments prohibit hurting animals on private property under their police power authority without provoking public controversy.[1]

Here we have an extension of the common law doctrine of police power to embrace members of the animal world other than humans that has raised no major argument. That is a notable development. Individual humans have nothing to gain from restrictions on hurting other animals beyond the peace of mind of knowing that it is not happening, so it is noteworthy that many governments use police power authority to institute these restrictions.

Hurting animals in many cases strikes an ethical nerve. That is surely because hurting humans strikes an ethical nerve, and we have rightly concluded there is no relevant difference between the two upon which to hang an ethical distinction. Pain is pain, whether animal or human. To behave as if it were otherwise would at best be factually mistaken and at worst, hypocritical.

Pain is a real phenomenon, albeit a complex and mysterious one. Each of us understands pain from personal experience, and we pay special attention to the attitude of others toward our pain when we decide what their status should be in our ethical community, just as we understand that we must pay appropriate attention to their pain. It is a classic quid pro quo. These social attitudes are associated with the capacity for empathy that we all expect of one another. That we accord this behavioral and cognitive significance to pain surely relates closely to its evolution as a publicly evinced phenomenon.

There can be no quid pro quo exchange with most other animals, and many animals even seem to be indifferent to the pain of others. Nevertheless, the principle of being sensitive to pain wherever it arises appears firmly ensconced in our ethical compact. To say the least, from a perspective of values there is something intrinsically urgent about responding to pain under ordinary circumstances. We have concluded that "pains" of comparable intensity and circumstances must be addressed alike as a matter of ethical conscience. Here some animals appear to be viewed as on a par with humans, ethically speaking, which is remarkable.[2]

It is also notable that this compassion does not help matters with regard to the fate of other species, or even the fate of our own. To the contrary, biologists know that compassion stands in the way of species health. Compassion for individual deer has led to herds far in excess of carrying capacity in many locations. Compassion for individual humans abounds worldwide even as the earth's population approaches eight billion, notwithstanding the stress to individuals and our species that this huge population ultimately entails.

Compassion rescues individuals, not species, because "species" cannot feel pain. The U.S. Fish and Wildlife Service would not have introduced wolves into the Greater Yellowstone ecosystem if its concern had been to minimize pain among individuals of the species that it oversees. Because of wolves, individual elk, moose, and other game have suffered that otherwise would not have suffered. It is even arguable that introducing wolves into that region has increased the sum total of pain in the world.

That suffering is inconsequential to the wildlife biologists because the forces of natural selection introduced through the wolves generate a variety of improvements at the level of the species and ecosystem, and these improvements are their goal. Similarly, improvements in species and ecosystems should be largely inconsequential to people whose sole concern is the fate of individual animals.

The truth is, for many people the specter of extinction of species, even our own, pales as an "ethical" concern in comparison to individual animal suffering. Prevention of animal suffering enjoys relatively uncontroversial police power protection that limits rights on private property, while preventing the extinction of species does not.

As the wildlife biologist James Estes observes,

People feel qualitatively different emotions about suffering individuals and suffering populations. Although many people are troubled intellectually by population declines or species extinction, these just don't seem to ring the same emotional chord as seeing an individual in distress. The consequences to human decision making are evident. Witness, for example, the disparate investments made by our society in improving human health versus controlling human population growth, despite the fact that human population growth is perhaps the gravest threat to a sustainable world.[3]

This is an anomalous situation, but perhaps it is not surprising. The locus of ethical deliberation is an individual deliberating person: you, me, and every other human being. In the framework of our rational self-interest, we are hence naturally programmed to respond to values that are seated within our consciousness as individuals, such as pain and pleasure, not to values at the level of our species, an abstract category whose relation to the interests of any individual is remote. The ethical compact embraces our identification with one another's pain relative to our own direct experience with and concern for pain as individuals. This is logically extended to pain among other individuals everywhere. There is no similarly simple way to arouse concern for species, even our own, from a logical extension of reflection upon our individual self-interest because none of us is a species or is fundamentally motivated as such.

To put it bluntly, a person may reasonably declare, "I am not the human species and, quite frankly, I do not particularly care about the human species in the long run any more than about any other species. I am a feeling, flesh-and-blood animal, and a species is nothing of the kind. I cannot even relate to the fates of species as compared with the fates of individual animals."

What's more, species cannot negotiate with us, cannot bargain with us to care about them, and this presents a problem for their standing in our ethics as well. We test and negotiate our ethical compact all the time, encounter by encounter, but we do so with individuals, not with species. There can be a quid pro quo to bind the terms of the ethical compact in these individual encounters. There cannot be a quid pro quo with species. Let us consider now the implications of this situation.

Social-contract theory as an account of the basis of ethics is convincing. The perspective of this theory is an age-old idea developed most brilliantly in modern times by John Rawls.[4] Social-contract theory explains social practices such as ethics on the analogy of agreements or contracts between individuals to achieve mutual advantages.

We do handle ethical negotiations as if they were contracts. But notice that this does not mean that social-contract theory implies an actual contract. Behavior in accordance with certain conventions abounds in the animal world, and this behavior follows from evolutionary developments that never required those conventions to be explicitly adopted.[5]

Bees that collect honey observe conventions of communication in displaying through movements to other bees the direction and distance of flowers.[6] The bees did not explicitly adopt those conventions. Nor did birds explicitly adopt the conventions of their calls to other birds, nor fish the conventions of their displays to other fish. The development of these many varieties of communication through evolution involved a great diversity of pathways through which the behaviors of the organisms came to be coordinated, and the creatures evolved along these different pathways without choosing them.[7]

Similarly, conventions of language pervade our own communication, but we never explicitly adopted them. That would have been impossible. A preexisting language would have been necessary in order to adopt such conventions explicitly. We speak to one another in accordance with conventions of grammar, but none of us can state all of them, and they are constantly in flux.

Ethical agreements may be likened to our conventions about grammar in that respect. In both cases we find ourselves early in life observing conventions of behavior without having expressly adopted them. But although we never expressly adopted those grammatical or ethical conventions, each of us knows how to discuss and enforce grammatical principles and ethical principles in order to maintain each system's viability, and we do so when necessary.

For example, truth telling is an ethical agreement that we never explicitly adopted. We would rarely say, "If you tell me the truth, I will tell you the truth." But a person who lies to us undercuts his or her own rightful expectation of truthful communication from us in return, and this fact is widely recognized. We may respond to someone who lies to us in some manner that will clarify the matter and resolve the breach, possibly in consultation with other people. Liars, therefore, risk jeopardizing their status within a larger ethical community than that of themselves and the person they lie to. That is how we test and negotiate our ethical compact all the time, encounter by encounter.

Now why should liars care about this loss of status? They are motivated fundamentally only to protect their own rational self-interest. How does status within an ethical community relate to rational self-interest?

They should care because, short of sociopathy or trying to dodge detection, they are better off embracing the cooperative principles of a community that accords them the same consideration they accord its members than they would be struggling alone outside the community. According to social-contract theory, those cooperative principles encompass "ethics."[8]

There is a profound difficulty in social-contract theory with regard to virtually all living things other than humans, as well as for all abstract categories like species, including the human species. These groups cannot engage with individual humans in regularly testing and negotiating the ethical compact. I develop and maintain my direct status in the ethical community, my "rights" therein, by my ethical behavior toward others and by enforcing my legitimate expectation to receive the same from them. These are not simple behaviors; they involve recursive series of mutual understandings. I know that you lied. I let you know that I know you lied. I know that you know that I know you lied. You know that I know that you know that I know you lied. And so on.[9] Recursive series of mutual understandings like this are implicit in all our communications; they are the essence of human communication.[10]

Animals that do not have the cognitive capacity to think and communicate in this way cannot enter directly into our ethical compact.[11] They can only gain indirect status within that compact, and only if we choose to confer it upon them. Any ethical rights that these animals receive through that compact do not arise as they do for us, directly by our skillful political enforcement of our legitimate standing. Rather, they arise by our extension of those rights without demanding a reciprocal obligation. Those animals can assert no independent claim on their own. They cannot be party to a contract.[12]

For this reason, assertions that many individual animals or abstract categories like "species" have "rights," or that they should be treated "fairly," are almost always absurd.[13] We might decide to protect "species" or individual animals under the law, as was done for the rights of endangered species under the U.S. Endangered Species Act, but that is not the immediate and direct way that rights arise and command authority among members of an ethical community.

Those particular animals and categories like species cannot even understand that we are according them rights. They cannot have any recognition of what constitutes "fair" or "unfair" behavior. To require fair treatment as a matter of ethics demands particular intellectual capacities. At a minimum, it requires a general ability to recognize when one is and when one is not being treated fairly, and when one's behavior toward another is fair or unfair. These concepts obtain their primary meaning and authority only in the hands of individuals asserting them in the quid pro quo of negotiation over terms of the ethical compact. It follows that social-contract theory largely excludes other forms of life from an ethical status comparable to our own.

Those of us who can and do engage in that negotiation thereby gain a kind of "intrinsic value" that plants, "cognitively incapable" animals, and abstract categories like "species" cannot gain. To assert that another person has no intrinsic ethical value for you destroys the ethical compact with that person. If someone asserts that you have no standing for him or her other than whatever use you may be to his or her own ends, you will make sure that that person is no longer a member of your ethical community.

The ethical compact demands a recognition of equal intrinsic ethical value among all the participants. It is an absolute condition of my being a member of an ethical community that I accord everyone else in the community a status that is equal to mine. That is the starting point of the compact, and it is the reason that other species, and indeed the human species as a species, are excluded from it.[14]

Why should any person, therefore, from the perspective of negotiating the social compact, choose to accord any other form of life ethical value? How does according another form of life ethical value further the public interest with regard to the ethical compact, which each of us has reason to protect against powerful motives like the acquisition of money?

The pace of loss of species on earth is accelerating. Much of the earth's living heritage remaining today may be lost forever. There may be little that we can do to reverse that process, so colossal are the opposing demographic and market forces. But there is no hope that countless forms of life can survive on this earth unless a consensus develops around a clear articulation and recognition of their value to us for what they are in themselves.

This cannot be the kind of intrinsic value that we have as participating members of an ethical community because other forms of life simply cannot achieve standing in that community the same way we can. It cannot be the value of "compassion" either, for this applies strictly to sentient organisms in certain circumstances and never to "species" as such. It also cannot be a monetary value which subordinates species protection to financial gain. But it must be an inherent value for us, not an instrumental value to serve our practical uses. Otherwise, forms of life that do not serve our practical uses will be lost, and that is a large percentage of the earth's species. What then, if anything, could such a value be? What realm of values have we missed?

EIGHT

What Kind of Humanity
Do We Embrace?

I magine a society unaffected by the inherent interest of other forms of life. This society watches indifferently as human activities extinguish species after species. Should we consider this society wrong?

To find fault presumes that survival of other species represents a greater value than a "matter of taste." But that value is exactly what this society denies. So if we want to protect species within this society, we are at a stand-off. Is there anything we can do?

Perhaps we could suggest an exchange: if the society will support species, we will support its own matters of taste. We make deals like this all the time with one another on a personal level: "You can have a dachshund, if I can have a boat." But such a procedure could not work in this case. Deals that are unsupported by shared values cannot protect many species from extinction. Population and market pressures on species today are too great.

We have come to a crossroads that demands that we make an intelligent decision. Protecting other living things for their own sake either matters to the public interest or it does not matter. Claiming that it matters is only so much hot air without a convincing and honest argument to back it up. What could that argument be?

It cannot be that other species have a conventional aesthetic value, as we have seen. It cannot be that they have values for our health, safety, survival, or recreation. It cannot be a statement of our values of compassion or rights or fairness. It cannot be tied to the values of the marketplace. These are all dead ends.

Clearly species on a vast scale command our protection for their uses to us. Many may figure in the carrying capacity of our planet for human survival. We try to patch together those uses and values with certain ethical claims, creating a hodgepodge of motivation that is helpful in protecting a species here and there. Yet for the countless species that do not serve those uses or merit those claims, such efforts seem contrived, unconvincing, and even desperate.

Worse, they beg the fundamental and decisive query that we

face: Do other living things have an inherent value to the public interest that merits the protection of humanity, or do they not? This is the question at the heart of our quest to save species.

As a practical fact, the answer may not matter, for there may be little hope at this point that we can avoid destroying a great many of the species on earth within the foreseeable future, whatever value we place in them. Our civilization does not seem to comprehend even now the full implications of our predicament.

But this much is certain: there can be no hope of saving other species of life on earth unless we at least understand why it matters to the public interest that we do so and how doing so ranks among competing values.

It would be nice if we could avoid the problem. It would be nice if we all agreed that it does matter whether other species are lost forever, considering what they inherently are, irrespective of our uses, and that it matters immensely.

But we do not all agree about that. Many people do not care about species. And so what? Indeed, what if many millions of people did not care about the fate of other life on earth? Well, millions don't. What difference does it make? Would they be better off if they did care?

Perhaps that is the real question.

Let us imagine a person who does not care about the fate of species, someone who is unaffected by the inherent interest of other forms of life. In fact, let us imagine someone who stands up and proclaims before all the world: "It does not matter if other species are lost forever to extinction."

Now, some might ask, "What kind of person would actually believe this? Where is the person who would stand up and proclaim such a thing before the world?"

Let us set aside trying to answer those two questions. Instead, let us reflect on the form of the questions themselves. The fact that we choose to phrase the inquiry in these distinctive ways is itself quite important; it points to an approach to understanding human values that has an ancient and noble lineage. We are ready now to explore the questions using this approach. Consider:

> What kind of person would not regret losing a species of life forever?
>
> Where can such a person be found?
>
> Where is the person who would not regret losing literature forever?
>
> What kind of person would not regret losing friendship?
>
> Where is the person who would not regret losing security for children? Losing justice? Losing democracy?
>
> What kind of person would not regret losing the canopy of stars?

We ask these questions in all seriousness. Each one of them provokes our solemn reflection. But not so the following questions:

What kind of person would not regret losing chocolate?

What kind of person would not regret losing cell phones?

Where is the person who would not regret losing tooth-

paste? Losing an umbrella? Losing air conditioning?

What kind of person would not regret losing television?

There is nothing solemn about these questions. They are awkward, if not humorous.

Why is the one set of questions serious and the other comical when their form is exactly the same? We all know why. We each mastered long ago the sophisticated skills in play here. These forms of question—"What kind of person . . . ?"/"Where is the person . . . ?"—are ones that we especially reserve to stimulate inferences and appraisals about a person's moral fiber, or broadly speaking, the principles by which that person lives. The result is amusing whenever the objective does not make sense: "What kind of person would not regret losing duct tape?"

Distinguishing among questions taking this form is nothing new for us. We do not find ourselves on unfamiliar turf. We are each adept at recognizing the qualities that shape the meaning in our lives. Questions like these are the way we distinguish qualities of character.

There are scores of terms in our language that reveal qualities of character, and they provoke our thinking and our communication with one another on a continuing basis: *courageous, para-*

noid, confident, callous, realistic, imaginative, curious, stubborn, honest, hopeful, mercenary, humble, grateful, apathetic, flippant, jealous, materialistic, mature, compassionate, hypocritical, friendly, obsessive, faithful, introverted, adventurous, foolhardy, reasonable, obsequious, reflective, cautious, wise, greedy, fair, gluttonous, optimistic, arrogant, patient, spiteful, consistent, reckless, responsible, dull, practical, narcissistic, sympathetic, indifferent, insightful, bigoted, hard-working, childish, spiritual, shallow, persistent, defensive, generous, and a host of others.

Some of these adjectives, like *fair* and *generous,* point to values that make immediate sense in the quid pro quo of social-contract theory. Other adjectives, such as *reckless, introverted,* and *shallow,* signal values that appear less related to social-contract bartering and have other implications. In effect, qualities of character trace an all-inclusive landscape of values that we find in human dispositions, and they amplify the choices we make that serve our rational self-interest. In other words, "rational self-interest" encompasses our preferences about character ("Would you rationally choose to be reckless?") as well as our preferences about specific things ("Would you rationally choose to sell your car today?").

Preferences about character seem to be superior to more quotidian preferences, for they tend to regulate how we behave in particular situations. When you reflect upon these qualities, you think of dispositions that manifest themselves throughout your behavior, much as a computer program operates with regard to diverse inputs for its outcomes. Your choosing to behave in a certain way in one situation (climbing a ladder on top of a chair on top of a table) and your choosing to behave in a certain way in a completely different situation (reading a map while shouting on the cell phone while driving through Chicago at 60 mph at rush hour) can both arise from a single trait of your character (being reckless).

Note that we ascribe exactly the same attribute to those two sets of behaviors, notwithstanding the entirely different actions and circumstances that they involve. Our estimation of both sets of behaviors stands or falls with our thinking of them as reckless—which like all character assessments entails a complex inference that we find ourselves making immediately. ("It was not reckless, because I was trying to save my child in both cases.")

We have here a new angle on our reasoning among values, and it turns out to be a profound one.[1] We are not neutral about questions of character in others or in ourselves. We reflect upon important matters when we ask, "What kind of person do I want to be?" Indeed, our judgment of what constitutes a life well lived rests entirely on our reasoning about qualities of perception and behavior that we seek for ourselves. Our reasoning about the flavors and commodities we like will not affect that judgment at all. These represent separate constellations of values that play on very different fields.

Appraisal of human qualities relies on a broad realm of values that find a compelling public reception. Appreciating aspects of character has a large payoff for our personal welfare and the long-term welfare of society alike. Choices of flavors and merchandise do not have such a payoff.

Indeed, admirable character traits cannot be bought or sold and remain either "admirable" or "character traits." We try to structure our institutions in the public interest to nourish attitudes that represent those values we embrace—being farsighted, being courageous, being realistic, being sincere.

We favor those attitudes as well by routinely analyzing and evaluating the behavior of others and ourselves: "What kind of person would do that?" In fact, it takes but a moment's reflection to realize that we seek to infer "character" from behavior all the time.

Some beliefs about character apply widely across our thinking and easily encompass our relationships to other forms of life. This distinguishes important beliefs about character from other ethical beliefs that resist extension to other forms of life, such as social-contract theory, in which the inability of other forms of life to join the ethical compact denies them a claim to its authority. When we think about qualities of character, our question shifts from whether other forms of life have direct rights under the ethical compact, which they cannot, to whether our best choices of character yield a decisive value for other forms of life independent of our uses.[2]

Thinking about values of character is different from appraising behavior strictly in terms of its causal outcomes. Thinking about character means appraising the frame of mind from which the behavior seems to emanate. The inquiry becomes one into the merit of the intentions that are manifested in our thought and conduct, an issue that we actually think about all the time.

It should come as no surprise that this revelation about our values is not new but ancient wisdom. Character analysis is the centerpiece of Aristotle's *Nicomachean Ethics,* in which Aristotle made paramount the question "What kind of person would do that?" He developed a systematic philosophy of human nature in which each person's faculty of "reason"—which he held to be a person's essence—would by nature seek attributes of character that optimize a flourishing life.[3] Aristotle called these attributes virtues.

Focusing on your personal fulfillment in a flourishing life provides a more accurate perspective on your "rational self-interest." The values that you favor are not measured only by the commodities that you buy with money but more fundamentally by the qualities of character you allow to influence what you buy. In other words, to gain a true sense of your values, it would be misleading to focus only on the specific things that you choose to do, for those choices may emanate from traits of your character that you would rather not have.

Aristotle's civilization was different from ours, and it flourished a long time ago, but a striking measure of his timeless insight is that the evaluation of qualities of character remains pervasive and fundamental in our reflections about ourselves and one another. Such evaluation is readily prompted by our attitudes about other species.

Here, for instance, is a characteristic remark about humans' indifference to species extinction that implies a criticism of the associated traits of character:

> The greatest tragedy in nature is the extinction of a species. . . . Where is the man who knowingly would stand by and watch a marvelous creation of nature—harmless to man's interests, and of no intrinsic commercial value—be forced into the vortex of extirpation without even raising his voice in protest?[4]

Notice that this comment is not about and could not be about the beauty of the creature or any analogous quality. Resting the comment on the shoulders of "beauty" would shift it from an appreciation centered in what we learn about the creature to one based in sensory and formal qualities that may be divorced from that learning. It would disenfranchise the first sentence and undermine the reference to commercial value: "The greatest tragedy in nature is the extinction of a species. . . . Where is the man who knowingly would stand by and watch a beautiful creation of nature—harmless to man's interests, and of no intrinsic commercial value—be forced into the vortex of extirpation without even raising his voice in protest?"

Beautiful in this context is like *flavorful* and seemingly extraneous to the public interest. Allusion to beauty would, in fact, destroy the rhetorical power of the comment. One would answer: "Where is that man? He is anywhere that the creature is not thought to be beautiful. Next question."

Marvelous, like *fascinating,* connotes a different set of values. We may disagree about beauty and flavors, but we rarely disagree with an observation that something is "marvelous" or "fascinating." Although initially we may resist such a designation, typically we tend to agree with it or to look into the "marvelous" thing more deeply for enlightenment (unless we are simply being argumentative and contrary). Further, in the final analysis, no one could reasonably suggest of *any* living thing that it is not a "marvelous" thing in all its extraordinary dimensions. Think of those dimensions. No honest defense of such a view is conceivable.

Beauty and flavor have connections to idiosyncratic physiology and cultural learning. Being "marvelous" or "fascinating" means having a compelling presence for the intellect and is arguably universal. A failure to appreciate what is marvelous or fascinating in our world is a failure to be curious, a failure to exercise thought.

Such a failure could be a matter of organic mental deficiency, or distraction, or even depression. But that a person might competently disparage curiosity about our world is patently absurd. It would be meaningless for a person to stand before any public forum professing such a position.[5]

A video game is marvelous. Einstein's theory of relativity is marvelous. Spiderwebs are marvelous. Each of these is a kind of technology. The first two are made by humans. The third is made by spiders.

> Where is the man who knowingly would stand by and watch a video game be forced into the vortex of extirpation without even raising his voice in protest?

This question is assuredly humorous. A video game is marvelous, but it is whimsical and self-referential, even frivolous. Further, since a video game is human-created, it lacks the mystery of the things in this world that are beyond our manufacture.[6] Its facets for curiosity are not substantially amplified by any connection to creation beyond itself, except for the psychology of human attention. That weakens immensely but does not completely eliminate its fascination.

The existence of a video game may be secure as a matter of market appeal, but even absent that the program for the game will presumably always exist or, if we save enough data, could be re-created. So any risk of extinction is probably unrealistic or easily averted.

> Where is the man who knowingly would stand by and watch Einstein's theory of relativity be forced into the vortex of extirpation without even raising his voice in protest?

This is a solemn question by any reckoning. Einstein's theory of relativity is marvelous, and it is not whimsical or self-referential, for it concerns the way the universe works. Its facets of interest are amplified by their connection to creation beyond itself, unless the theory some day becomes implausible and is replaced. That would greatly weaken its fascination. Otherwise, a more compelling object of curiosity is not imaginable. As in the case of the video game, however, any risk of extinction is unrealistic, for the theory has entered our way of viewing the universe and could be generated again from human reasoning.

> Where is the man who knowingly would stand by and watch spiderwebs be forced into the vortex of extirpation without even raising his voice in protest?

This question has a completely different impact. Spiderwebs are marvelous, and they are not whimsical or self-referential, for they connect to all the myriad dynamics of evolution spanning untold eons of time. Among other facets of spiderwebs, they are necessary for the existence of most spiders. Consequently, spiders—living things with their profound dimensions of significance—will be lost if their webs are lost. These dimensions include astonishing elements of a spider's design that have arisen from unique and unrepeatable flows of adaptation through endless environments in the earth's career. They include all the relationships across a living geography to the forms of life ecologically connected with spiders.

This barely touches the depths of interest of spiders for us as forms of life. In particular, we are living things as well as they, so interest in spiders as living things has a profound personal connection to the human perspective. We share with spiders the same primordial origins on this earth. We share the same vast tree of genealogy. We share with them the same dimensions as forms of life. We share with spiders the same epochs of adaptation and survival, and even the same basic constraints of fate.

Einstein's theory of relativity is *about* elements of creation, whereas spiderwebs *are* elements of creation, and their destruction would preclude forever further enlightenment about spiders and all the life in this world that is related to them. Whereas Einstein's theory will presumably always exist, or could be replicated, so that risk of its extinction is far-fetched, extinction of species of spiders and their webs in the wake of human activities is a genuine likelihood. Any such extinction would terminate permanently and irrevocably this example of a continuity of life alongside our own that has been part of the earth's unfathomable flow of impacts over billions of years of evolution.

Imagine that a species of spider is nearing extinction from the encroachment of human activities through the last of its range, which happens to be in the United States. The spider has no value in any market. Now imagine that sales of a certain video game explode. The video-game company leases a large tract of land to expand sales and production. A pivotal population of the spider is discovered living on this land. The U.S. Fish and Wildlife Service blocks activity under the Endangered Species Act. Either the company or the landowner thereby loses the entire value obligated under contract for the duration of the lease. The value of the property, including any chance of a subsequent lease, plummets. The hope of jobs and an enhanced tax base for the region dims.

One or the other must go. It is either survival for this species of spider or survival for the building, the property value, the jobs, the economic effects. This is not an isolated case. As market-driven commerce expands worldwide, species after species vanishes.

The human species is not at risk in this situation, and no human life is directly at risk, either. Despite a pumped-up market, the video games are a trivial amusement, and their basic programs are not threatened. The company could probably make other arrangements. People will lose money, but people lose money from failed investments all the time. In other words, extinction of the species represents an absolute and irrevocable loss of a unique living heritage from our planet, while prevention of the building is a matter of temporary inconveniences.

It is such a simple choice for the public interest.

This is not to deny the great hardships connected with the economic adjustments that must be made in many cases like this. But hardships have accompanied the protection of the public interest from time immemorial, and they can be resolved only through ingenuity, with faith in the importance of the outcome. The issue of unfairness and inequality in the way these sacrifices are shared among the people affected must be addressed, but that should never be confused with the issue of safeguarding the public interest for future generations. Pitting the survival of species against human welfare in a given instance as if there were no other option presents a false dilemma that is hastening the destruction of our planet. In every case, we are weighing the plenitude of the earth's wonders available to our descendants against changing our ways of living to respect our approaching limits.

This does not mean that construction of every building must be halted and every species saved. That may not be realistic. It just means we must accept that other living things are as profoundly fascinating and deeply personal for human mindfulness as anything under creation, and that in virtually every case our choice between the survival of a species and material benefits is obvious.

We are far from this acceptance when people react with contempt to the notion, or with anger at the species. How wretched to respond by being angry at the spider, or a bird or a frog. Humans do, indeed, perhaps as a matter of our nature, compete personally and fiercely with other forms of life, even seeking domination over them. Perhaps this behavior had evolutionary advantages for us at one time. But this kind of display today is misplaced and even ludicrous, much like a clown caught up in feuding with puppets and infants.

Relating to other forms of life in terms of our gut projections instead of as they actually are is intellectually dishonest. A spider does not have anything to do with those projections. It is a staggering phenomenon billions of years in the making whose full story we can scarcely glimpse. It is an artifact of untold ages of earth environments, just as we are of those same ages. It is not "ugly." It is not "to blame." Much like us, a spider is a consummately marvelous form of life, carrying forward through its brief and extraordinary existence a flame of life that is almost as old as our planet.

That is what a spider, a bird, and a frog actually represent. That is what every living species on earth represents. What kind of person would turn his or her back upon such a thing?

Let us suppose that societies worldwide permit all the buildings to be built in all these situations, and that then, with increasing population and further market expansion, more buildings are built, and then still more. Let us suppose that by some miracle all the unknown carrying capacities relative to sustaining human populations on earth become precisely known. The final brakes must be applied to the expansion just in time, at those points of carrying capacity, to rescue humanity from catastrophe.

Now it is a mystery how humanity will apply the final brakes just in time, if humanity cannot apply them now, for it will be much harder to do so then.

Further, none of us individuals who are negotiating the social contract are "species," nor are our children or grandchildren "species." So how will our rational self-interest motivate enough concern about the human species for us to put the brakes on, particularly in an age of all-consuming market pressures?

The following reasoning from individual self-interest surely circulates at some level of consciousness even now:

> Why should I care about the fate of my own species, for it cannot make me happier or rescue me from death, and I will be long dead before its fate is decided?

But let us suppose that the final brakes are miraculously applied at the relevant points of human carrying capacity. In those desperate times, the same temperament of restraint that was sidestepped before will be imperative now, but magnified many times over, and with little margin for choices. The human predicament will genuinely be dire, and our adjustments will have to be draconian. The impacts on values in every corner of our lives will be grim.

In other words, from the standpoint of the public interest, permitting that first video-game factory at the expense of the spider will not in the end prevent any misfortunes at all. It will merely delay all the misfortunes until the day we finally have to put the brakes on. Meanwhile, the amazing creature that traces an extraordinary living kinship with us to the very cradle of life on this planet will be gone forever, and for nothing. The building that destroyed the species will only postpone the human consequences until a day when they become immensely worse.

We are needlessly losing forever the legacy of life on earth through such behavior. This disappearing heritage includes connections to our own identities in evolved, earthbound biodiversity. We are transforming this world into endlessly duplicated, mind-numbing monocultures of human-serving products. We are erasing the variety of life that distinguishes the earth among all known planets in the universe.

We are awash in these devastating consequences because we are failing as a civilization to follow our own grade-school admonition, one of our golden rules of character, and putting off until tomorrow problems we have to solve today.

To put it bluntly, the effect of humans on this planet could conceivably be much like that of locusts in a field of grain. But the effect of locusts does not matter to any awareness of which locusts are capable, while the effect of humans does matter to the awareness of humans.

What more is there to say? Here is an opportunity and an irony perhaps never to be repeated in the universe, that a species of life has evolved with the ability to contemplate in wonder the miracle of life itself, and is knowingly engaging in the destruction of that miracle. Our own seeming defects of reason and motivation trap us in a limited range of pursuits that maximize market-driven results for short-term gratification over character-driven results for long-term fulfillment.

Like it or not, here we are. Even as our singular human character responds to the infinite marvel of living creation that is all around us, other aspects of our nature propel us to annihilate that creation. Lending a more effective voice to the realms of value of all living species for a flourishing human spirit is essential to securing any saner outcome.

The loss of frontiers on earth, the pressures on its carrying capacities—these present us with a dilemma of proportions never before faced in human history. It was only in the eighteenth century that Linnaeus replaced the chaos of using myriad languages and locally varying common names to identify living things with a uniform classification that could be understood throughout the world. Before everyone had a consistent name to call a particular species there was in many instances no way to know that it was disappearing, because we could not have counted its population. It was not until Charles Darwin and Alfred Wallace in the second half of the nineteenth century that we were able to realize the fascinating evolutionary history inherent in each species, and to understand that extinction is forever.

Now, in the twenty-first century, it is being brought home all across the world that ours is a finite planet, and that the first casualties of our human excesses are the planet's diversity of species, billions of years in the creation.

Suppose that we turn our heads away. Suppose we ignore this decline in our planet's abundance and proceed as usual. Today this may mean simply that one more resource (a species of fish, for example, or a particular kind of mineral) is no longer affordable or is removed from the marketplace altogether because the earth's supply is exhausted. But one by one, in time to follow, more and more things that money can now buy will likewise disappear. The values that money could never buy, such as life itself in every astonishing form, will have been increasingly decimated as well, and to no end but a futility of delay and procrastination.

Why should we care? We will not be around. It will not affect us.

This is a decent question for explicit and honest attention, and if we ignore it, then decimation and decline will be our answer. One way or the other, our actions will play out to consequences. So, again, why should we care? Why should any of us sacrifice anything today for future human beings whom we will never know? In such a sharp light as this, it seems that there may be very limited reason to care about our own species or any other.

But perhaps the lure of the marketplace has pulled us from our sanity, and we have ceased to reason well about our reasons, and can no longer discern the true measure of our human spirit. Perhaps day by day, "what we have reason to care about" is losing its voice in the noble landscape where higher values modulate our "reasons," and where ideation about our deepest commitments necessarily transcends matters of our lifespan.

This is a landscape in which values arise from our qualities of character, and thoughtful hands are placed on the controls. It is a landscape of sure values, grounded in a deepening knowledge of timeless creation that casts the veneer of civilization in lesser and truer proportions, and in which, if we are honest and curious, we can at last begin to see that a failure to appreciate other species of life on earth is a failure to appreciate ourselves.

NOTES

ONE To an Inquisitive Mind Open
to Honest Reflection, the Value of Every
Species Is Incalculable

1. Robert B. May, "Ecological Science and Tomorrow's World," *Philosophical Transactions of the Royal Society B: Biological Sciences* 365 (2010): 41–42. See also American Museum of Natural History, "National Survey Reveals Biodiversity Crisis—Scientific Experts Believe We Are in the Midst of the Fastest Mass Extinction in Earth's History," news release, April 20, 1998; Edward O. Wilson, *The Future of Life* (New York: Knopf, 2002). The International Union for the Conservation of Nature and Natural Resources (IUCN) reports that more than a third of 47,677 assessed species worldwide are threatened with

extinction. (IUCN, "Extinction Crisis Continues Apace," press release, November 3, 2009). A study by more than a hundred scientists the following year found that "extinctions have increased to between 100 and 1,000 times greater than the rate they were in the distant past": J. E. M. Baillie et al., eds., *Evolution Lost: Status and Trends of the World's Vertebrates* (London: Zoological Society of London, 2010), 64.

2. Jennifer B. Hughes, Gretchen C. Daily, and Paul R. Ehrlich, "Population Diversity: Its Extent and Extinction," *Science* (October 24, 1997): 689; Paul R. Ehrlich and Anne H. Ehrlich, *Extinction* (New York: Ballantine, 1981), 36–38; Mark L. Schaffer, "Minimum Population Sizes for Species Conservation," *BioScience* 31(2) (February 1981): 131.

3. See *Millennium Ecosystem Assessment Series, Ecosystems and Human Well-Being: General Synthesis* (Washington, D.C.: Island, 2005).

4. Ehrlich and Ehrlich, *Extinction*, xi. Extinguishing species and populations is a perilous gamble; see Paul Ehrlich and Brian Walker, "Rivets and Redundancy," *BioScience* 48(5) (May 1998): 387–88.

5. See Leopold's classic "The Land Ethic," in Aldo Leopold, *A Sand County Almanac: With Other Essays on Conservation from Round River* (New York: Oxford University Press, 1966), 217. Leopold writes, "All ethics so far evolved rest upon a single premise: that the individual is a member of a community of interdependent parts. . . . The land ethic simply enlarges the boundaries of the community to include soils, waters, plants, and animals, or collectively, the land" (219).

6. These are issues that plainly affected Leopold's perspective. See Julianne Lutz Newton, *Aldo Leopold's Odyssey: Rediscovering the Author of "A Sand County Almanac"* (Washington, D.C.: Island, 2006), 340–41. Newton writes of Leopold's conservation philosophy: "Integrity had to do with the parts of nature, but it meant more particularly the parts of nature *that were necessary for land to keep its sta-*

bility and health. But what parts were necessary?" (340, emphasis added).

7. See "Biodiversity," *Africa Environment Outlook 2: Our Environment, Our Wealth* (Nairobi: United Nations Environment Programme, 2006), 226–61.

8. For discussions of the distinction see, with associated references, Andrew Light and Holmes Rolston III, eds., *Environmental Ethics: An Anthology* (Oxford: Blackwell, 2003): 129–90; Bryan G. Norton, *Searching for Sustainability: Interdisciplinary Essays in the Philosophy of Conservation Biology* (Cambridge: Cambridge University Press, 2002); and Wayne Ouderkirk and Jim Hill, eds., *Land, Value, Community: Callicott and Environmental Philosophy* (Albany: State University of New York Press, 2002). See also Chapters 7 and 8, below. Intellectual values in the qualities of character that we revere are examined in Michael DePaul and Linda Zagzebski, eds., *Intellectual Virtue: Perspectives from Ethics and Epistemology* (New York: Oxford University Press, 2003).

9. *Encyclopaedia Britannica: Micropedia*, 15th ed., s.v. "aesthetics."

10. Alternative conceptions of aesthetics may escape this problem through closer alignment with arguments like those that follow regarding the intellectual appreciation of an object in itself. The same challenge may confront Edward O. Wilson's compelling "biophilia hypothesis." This theory proposes that evolution has produced within humans innate bonds of feeling for other life. See Edward O. Wilson, *Biophilia* (Cambridge: Harvard University Press, 1984). Even if that were so, these innate affections would be nonrational impulses that might vary from one person to another and across forms of life as objects of affection, which would compromise their utility as a rational basis for apprehending the inherent value of living things.

11. For a magnificent account of life on earth, see Edward O. Wilson, *The Diversity of Life* (Cambridge: Harvard University Press, 1992).

12. For an exploration of our earliest ancestors, see J. William Schopf, *Cradle of Life: The Discovery of Earth's Earliest Fossils* (Princeton: Princeton University Press, 2000).

13. I owe this example to Charles Jones.

14. I owe this perspective to Gary Beauvais.

15. John A. Byers, *American Pronghorn: Social Adaptations and the Ghosts of Predators Past* (Chicago: University of Chicago Press, 1997), 11.

16. I am indebted to Gary Beauvais and Charles Jones for some of the insights in this paragraph.

17. For a panoramic perspective on this vast transmission of cultural versus genetic information and its grave consequences today, see Paul R. Ehrlich and Anne H. Ehrlich, *The Dominant Animal: Human Evolution and the Environment* (Washington, D.C.: Island, 2008); and Paul R. Ehrlich, *Human Natures: Genes, Cultures and the Human Prospect* (Washington, D.C.: Island, 2000).

18. I owe this imagery to Thomas Lovejoy. See his essay "Biological Diversity" in *Life Stories,* ed. Heather Newbold (Berkeley: University of California Press, 2000), 42–54. Lovejoy explains, "One irreplaceable value of nature is as a living library on which to build the life sciences. Each species is a unique set of solutions to a specific set of biological problems, equivalent not to a book but to a series of volumes. If we lose a species we lose that knowledge. Unlike information in books, once the species is gone, all the information goes with it"(53).

19. This is roughly the range of dates agreed on in studies of speciation rates. See Jerry A. Coyne and H. Allen Orr, *Speciation* (Sunderland, Mass.: Sinauer Associates, 2004), 411–45.

TWO The Intellectual Value of
Species to Humans Stems from Our
Unique Character

1. I use *creation* to mean "existence" and do not mean to imply a sentient creator.

2. George Loewenstein, "The Psychology of Curiosity: A Review and Reinterpretation," in *Exotic Preferences: Behavioral Economics and Human Motivation,* ed. George Loewenstein (New York: Oxford University Press, 2007), 121, 169.

3. Susan Haack, "The Ideal of Intellectual Integrity, in Life and Literature," *New Literary History* 36 (2005): 359, 360, 365. This is not to deny that someone may learn not to inquire into things, as Harold Rollins observed to me. Alternatively, someone may be so involved with personal challenges or supernatural beliefs as to not be curious about the natural world.

4. Philosophers, psychologists, anthropologists and archaeologists explore the evolution of mental attributes in Peter Carruthers and Andrew Chamberlain, eds., *Evolution and the Human Mind* (Cambridge: Cambridge University Press, 2002).

5. Indeed, some studies even suggest that an unrelenting barrage of urban versus "natural" stimuli may diminish cognitive functioning. See Mark G. Berman, John Jonides, and Stephen Kaplan, "The Cognitive Benefits of Interacting with Nature," *Psychological Science* 19(12) (2008): 1207–12.

6. Martin Antony and Randi McCabe, *Overcoming Animal and Insect Phobias: How to Conquer Fear of Dogs, Snakes, Rodents, Bees, Spiders and More* (Oakland, Calif.: New Harbinger, 2005), 10.

7. See Charles Barsotti's cartoon of two lions conversing: "Right. How Could Anyone Look at a Rotting Zebra Corpse and Not Believe There's a God?" *New Yorker* (January 10, 2000).

8. For diverse perspectives on anthropomorphism by scholars from many disciplines, see Robert Mitchell, Nicholas Thompson, and H. Lyn Miles, eds., *Anthropomorphism, Anecdotes, and Animals* (Albany: State University of New York Press, 1997).

9. See Donald Kroodsma, *The Singing Life of Birds: The Art and Science of Listening to Birdsong* (Boston: Houghton Mifflin, 2005).

10. Anthropomorphic reactions do present challenges for studies of animal behavior. This is discussed in John S. Kennedy, *The New Anthropomorphism* (Cambridge: Cambridge University Press, 2003).

11. According to the child psychologist and neuroscientist Colwyn Trevarthen, "The nature of human learning is unique. It is driven by a different kind of curiosity than that which other animals possess. It is both imitative and creative and makes greater use of communication"; Trevarthen, "The Child's Need to Learn a Culture," in *Cultural Worlds of Early Childhood,* ed. Martin Woodhead, Dorothy Faulkner, and Karen Littleton (London: Routledge, 1998), 97.

12. For a discussion of methodology in semantics that involves "the use of truth as the central semantic concept," see Michael Luntley, *Contemporary Philosophy of Thought: Truth, World, Content* (Oxford: Blackwell, 1999): 104–5.

13. It is true that supernatural beliefs acquired in later childhood and adolescence in cultures throughout the world can deflect this native curiosity. Supernatural beliefs survive through social sanctions that restrict active curiosity. Supernatural beliefs about other living things are not about living things as they are in themselves because such beliefs discount empirical evidence. Supernatural beliefs that obstruct

curiosity and intellectual honesty can present serious challenges to the appreciation of the inherent value of living things. For a discussion of the evolution of supernatural belief systems, see Candace S. Alcorta and Richard Sosis, "Ritual, Emotion, and Sacred Symbols," *Human Nature* 16 (Winter 2005): 323–59.

T H R E E   The Fate of Life on Earth
Hinges on Property Values

1. Of course, this distinction between owning a species and owning an individual organism becomes meaningless as species become rare. See Holmes Rolston III, "Property Rights and Endangered Species," *University of Colorado Law Review* 61 (1990): 283, 293.
2. No other species in the world has ever spread its population so widely or upset life so extensively as humans.
3. *Commonwealth v. Alger,* 61 Mass. 53, 84–85 (1851).
4. The United States Supreme Court discussed these matters and discarded this formula in *Lucas v. South Carolina Coastal Council,* 505 U.S. 1003, 1025–26 (1992).
5. If the government could afford the judicial administration of claims and the cost of reimbursing landowners under legislation like the Endangered Species Act, such controversial legislation would arguably not be subsumed under police power.
6. See Mark Sagoff, "Muddle or Muddle Through? Takings Jurisprudence Meets the Endangered Species Act," *William and Mary Law Review* 38 (1997): 825. For a proponent of property rights over species protection, see James W. Ely, Jr., "Property Rights and Environmental Regulation: The Case for Compensation," *Harvard Journal of Law and Public Policy* 28 (2004): 51.

7. To make matters worse, private property rights can have impacts much broader than the effect of an action on a single parcel of land. Ecological effects spread far beyond a particular property through predator-prey interdependencies, species migration patterns, seed dispersals, and other phenomena.

8. A popular bumper sticker in Oregon reads, "Spotted owl tastes like chicken." In Florida, people are advised, "Save a logger. Eat a woodpecker." In the 1992 campaign against Bill Clinton, George H. W. Bush mocked Al Gore as "Ozone Man" and claimed, "This guy is so far out in the environmental extreme we'll be up to our necks in owls and outta work for every American" (David Remnick, "Ozone Man," *New Yorker* [April 24, 2006]). See Steven L. Yaffee, "The Northern Spotted Owl: An Indicator of the Importance of Sociopolitical Context," in *Endangered Species Recovery,* ed. Tim W. Clark, Richard P. Reading, and Alice L. Clarke (Washington, D.C.: Island, 1994).

FOUR   Humans Are Poised to Destroy
the Resources of a World of
Bountiful Interest

1. This starting point may consistently lead a person to deny his or her own interests for the sake of service to another cause that he or she implicitly or explicitly finds reason to embrace.

2. This observation is consistent with a conclusion that humans fundamentally need the love of other humans as a general fact of human psychology and health.

3. Garrett Hardin, "The Tragedy of the Commons," *Science* (1968): 1243–48. For recent analysis of this concept and its applications, see John A. Baden and Douglas S. Noonan, eds., *Managing the Commons* (Bloom-

ington: Indiana University Press, 1998); Donald Kennedy, ed., *Science Magazine's State of the Planet, 2006–2007* (Washington, D.C.: Island, 2006): 101–93; and Nives Dolšak and Elinor Ostrom, eds., *The Commons in the New Millennium: Challenges and Adaptation* (Cambridge: MIT Press, 2003).

4. Charles Darwin explained elegantly why brutal self-interest was a part of natural selection. Any individual who holds back is knocked out. It should not be surprising that our practical reasoning largely exhibits this same priority of self-interest. However, there are also Darwinian accounts of the evolution of reciprocal altruism and kin selection in the field of sociobiology that would explain departures from individual self-interest in human behavior. But these mechanisms clearly offer little to counteract the innumerable examples of the tragedy of the commons occurring throughout the world.

5. A broader formulation of this theory is known as Coase's theorem. The economist Ronald Coase demonstrated in a classic article that if property rights were assigned clearly and there was no cost in negotiating for their use, then all allocations of property would avoid economically inefficient results like the tragedy of the commons because the interested parties would bargain according to self-interest to ensure efficient results. See Ronald Coase, "The Problem of Social Cost," *Journal of Law and Economics* 3 (1960): 1–44.

6. Ulysses S. Seal, E. Tom Thorne, Michael A. Bogan, and Stanley H. Anderson, eds., *Conservation Biology and the Black-Footed Ferret* (New Haven: Yale University Press, 1989).

7. Randal O'Toole, "The Tragedy of the Scenic Commons," in *Managing the Commons,* ed. John A. Baden and Douglas S. Noonan (Bloomington: Indiana University Press, 1998), 181–87.

8. At least, one comes to look at things this way in a commodified world-

view. See Margaret Jane Radin, *Contested Commodities* (Cambridge: Harvard University Press, 1996). See also Chapter 6 below, and the references at notes 4 and 11.

FIVE Property Ownership and the Desire for Money Work Against the Interests of Species

1. Illegal trade in endangered species throughout the world is chronicled in the TRAFFIC Bulletin published in Cambridge. TRAFFIC, the Wildlife Trade Monitoring Network, is a joint program of the World Wildlife Fund and the International Union for the Conservation of Nature.

2. For trenchant research on this bias, see George Loewenstein, Daniel Read, and Roy F. Baumeister, eds., *Time and Decision: Economic and Psychological Perspectives on Intertemporal Choice* (New York: Russell Sage, 2003). The primatologist Hans Kummer observes that short-term gratification under evolution must actually serve long-term survival, but only so long as the surrounding conditions do not materially change from those that evolution prepared for: "It is astonishing that gratification in this sense really should be a reliable pointer to long-term survival. This highly improbable correlation in animals under natural environmental conditions is the achievement of evolution. In the long run, selection does not permit any organization of behavior in which actions with high gratification value have a low survival value. . . . A more reliable principle would no doubt be for the choice of behavior to depend directly on its survival value, but that would require brains capable of a degree of insight that not even humans possess, to say nothing of more primitive life forms. The evo-

lution of life began with microorganisms having no brain at all and therefore could not have proceeded according to the exacting principle of insight" (Hans Kummer, *In Quest of the Sacred Baboon: A Scientist's Journey* [Princeton: Princeton University Press, 1997], 140).

3. Gary Beauvais called this example to my attention. See "Grizzly Bear Survival May Depend on Fragile Moths," *American Scientist* (May–June 2002), 295.

4. Sandra L. Olsen, ed., *Horses Through Time* (Lanham, Md.: Roberts Rinehart, 1996).

5. Mashbat Sarlagtay, "Mongolia: Managing the Transition from Nomadic to Settled Culture," in *The Asia-Pacific: A Region in Transition,* ed. James Rolfe (Honolulu: Asia-Pacific Center for Security Studies, 2004), 323.

6. Ibid., 324.

7. But not long ago American farmers, cowboys, and shepherds fighting range wars tangled with this idea of liberty themselves, which Cole Porter immortalized in his song "Don't Fence Me In." Traces of this sentiment, which was shared by Native Americans, can still be found in the American West.

8. Morris Rossabi, *Modern Mongolia: From Khans to Commissars to Capitalists* (Berkeley: University of California Press, 2005). The Constitution of Mongolia, Chapter 1, Article 5, provides: "1. Mongolia shall have an economy based on different forms of property and answering both universal trends of world economic development and national specifics. 2. The State recognizes all forms of both public and private property and shall protect the rights of the owner by law."

9. These land assets amount to trillions of dollars of untapped assets. See Hernando de Soto, *The Mystery of Capital: Why Capitalism Triumphs in the West and Fails Everywhere Else* (New York: Basic Books, 2003),

and the review of it by Richard Pipes, "The Mystery of Capital by Hernando de Soto," *Commentary* (January 2001), 65–68.

10. These are the sorts of rationales that the World Bank and others offer to encourage nations such as Mongolia to privatize resources. See Tim Hanstad and Jennifer Duncan, "Land Reform in Mongolia: Observations and Recommendations," *RDI Reports on Foreign Aid and Development* (Seattle: Rural Development Institute, 2001), 13–16.

11. For a discussion of land privatization in Mongolia, see Chinzorig Batbileg, "Does Land Privatization Support the Development of a Land Market?" paper presented at the International Workshop: Land Policies, Land Registration and Economic Development, Experiences in Central Asian Countries, Tashkent, Uzbekistan, October 31– November 3, 2007.

12. The Constitution of Mongolia, Chapter 1, Article 6, provides: "3. The State may give for private ownership plots of land except pastures and areas under public and special use, only to the citizens of Mongolia."

13. Orhon Myadar, "Nomads in a Fenced Land: Land Reform in Post-Socialist Mongolia," *Asian-Pacific Law and Policy Journal* 11 (2009): 161, 174, 185–86, 191.

14. Ibid., 195–203.

15. Naranchimeg Bagdai et al., "Transparency as a Solution for Uncertainty in Land Privatization—A Pilot Study for Mongolia," Proceedings of the FIG Working Week: Surveyors' Key Role in Accelerated Development, Eilat, Israel, May 3–8, 2009, 8.

16. The Constitution of Mongolia, Chapter 2, Article 16, provides: "The citizens of Mongolia are guaranteed to enjoy the following rights and freedoms: ... 3) The right to fair acquisition, possession and inheritance of movable and immovable property. Illegal confiscation and requisitioning of the private property of citizens shall be prohibited.

If the State and its bodies appropriate private property on the basis of exclusive public need, they shall do so with due compensation and payment."

17. The Constitution of Mongolia, Chapter 1, Article 6, provides: "1) The land, its subsoil, forests, water, fauna and flora and other natural resources shall be subject to national sovereignty and State protection." Further, it provides in Chapter 2, Article 16: "The citizens of Mongolia are guaranteed to enjoy the following rights and freedoms: . . . 2) The right to healthy and safe environment, and to be protected against environmental pollution and ecological imbalance."

18. Toward this enforcement in Mongolia, the key role of "transparency" in privatization is discussed in Bagdai et al., "Transparency as a Solution for Uncertainty in Land Privatization."

19. We must explicitly renew our values with every generation if we hope to save our own or any other species. This is another profoundly challenging aspect of our predicament with human nature.

S I X   Free Market Environmentalism
Places Profits Above the
Public Interest

1. I am indebted to Thomas Power for stimulating this critical analysis of unbridled free market environmentalism. Thomas M. Power, "Ideology, Wishful Thinking, and Pragmatic Reform," in *The Next West*, ed., John Baden and Donald Snow (Washington, D.C.: Island, 1997): 233.

2. Clean Air Act Amendments of 1990, 42 U.S.C. 7401 et seq.

3. The Clean Air Act emissions-trading program has exceeded expectations in cost-effective pollution reduction. See Dallas Burtraw and Karen Palmer, "The Paparazzi Take a Look at a Living Legend: The

SO$_2$ Cap-and-Trade Program for Power Plants in the United States," Discussion Paper 03-15, Resources for the Future, Washington, D.C. (April 2003); Richard A. Kerr, "Acid Rain Control: Success on the Cheap," *Science* (November 6, 1998): 1024.

4. See Power, "Ideology, Wishful Thinking, and Pragmatic Reform," 242–43. For a penetrating analysis of money, commodities, and non-monetary values, see chapter 4 of Michael Walzer's classic *Spheres of Justice: A Defense of Pluralism and Equality* (New York: Basic Books, 1984), 95–128, and succeeding commentary in David Miller and Michael Walzer, eds., *Pluralism, Justice and Equality* (New York: Oxford University Press, 1995). See also chapter 5 of Amartya Sen, *Development as Freedom* (New York: Anchor, 1999), 111–45, and the references below at note 11.

5. Terry L. Anderson, "Markets and the Environment: Friends or Foes?" *Case Western Reserve Law Review* 55 (2004): 81–91.

6. As one commentator puts it: "In that event, the economist should view with equanimity the felling of ancient forests to ship raw logs to Japan, the ownership of Yellowstone by the Walt Disney Company or the transfer of the water in their favorite streams and rivers to suburban lawns in Las Vegas and Phoenix" (Mark Sagoff, "Saving the Marketplace from the Market," in Baden and Snow, *Next West,* 145).

7. Terry L. Anderson and Donald R. Leal, *Free Market Environmentalism* (New York: Palgrave, 2001).

8. Michael D. Copeland, "The New Resource Economics," in *The Yellowstone Primer: Land and Resource Management in the Greater Yellowstone Ecosystem,* ed. John Baden and Donald Leal (San Francisco: Pacific Resource Institute for Public Policy, 1990), 16.

9. Donald Snow, "Empire or Homelands? A Revival of Jeffersonian Democracy in the American West," in Baden and Snow, *Next West,* 181–203.

10. Mark Barringer, *Selling Yellowstone* (Lawrence: University of Kansas Press, 2002), 34.

11. Among the recent scholarship on commodification and its boundaries, see Martha Ertman and Joan Williams, *Rethinking Commodification: Cases and Reading in Law and Culture* (New York: New York University Press, 2005); Susan Strasser, *Commodifying Everything: Relationships of the Market* (New York: Routledge, 2003); and Margaret Jane Radin, *Contested Commodities* (Cambridge: Harvard University Press, 1996), 48.

12. Wilderness Act of 1964, 16 U.S.C. 1131 et seq.

13. Power, "Ideology, Wishful Thinking, and Pragmatic Reform," 239.

14. Aldo Leopold reduced this claim to dramatic numbers. "One basic weakness in a conservation system based wholly on economic motives," he wrote, "is that most members of the land community have no economic value. Wildflowers and songbirds are examples. Of the 22,000 higher plants and animals native to Wisconsin, it is doubtful whether more than 5 percent can be sold, fed, eaten, or otherwise put to economic use" (Leopold, *A Sand County Almanac: With Other Essays on Conservation from Round River* [New York: Oxford University Press, 1966], 225).

15. Michael Grunwald, *The Swamp* (New York: Simon and Schuster, 2006), 4; Marjory Stoneman Douglas, *The Everglades: River of Grass* (New York: Rinehart, 1947), 5.

SEVEN Species Have No Direct
Claim for Consideration in
an Ethical Community

1. See Margit Livingston, "Desecrating the Ark: Animal Abuse and the Law's Role in Prevention," *Iowa Law Review* 87 (2001): 1–73.

2. For a discussion of animal pain and the law, and the difficult issues in identifying and assessing pain among animals, see Colin Allen, "Animal Pain," *Noûs* 38 (2004): 617–43.

3. J. A. Estes, "Concerns About Rehabilitation of Oiled Wildlife," *Conservation Biology* 12 (1998): 1157.

4. John Rawls, *A Theory of Justice,* rev. ed. (Cambridge: Belknap, 1999).

5. The seminal analysis of this broadly adaptive concept of a "convention" is David K. Lewis, *Convention: A Philosophical Study* (Cambridge: Harvard University Press, 1969).

6. Karl Von Frisch, *The Dance Language and Orientation of Bees* (Cambridge: Harvard University Press, 1967).

7. Evolution of animal communication is the subject of John Maynard-Smith and David Harper's *Animal Signals* (New York: Oxford University Press, 2004).

8. Rawls, *Theory of Justice,* 4.

9. Alasdair MacIntyre, "Truthfulness and Lies: What Can We Learn from Kant?" *Ethics and Politics: Selected Essays* (New York: Cambridge University Press, 2006): 141.

10. Paul Grice, *Studies in the Way of Words* (Cambridge: Harvard University Press, 1989), 213–23; Lewis, *Convention,* 24–33, 152–56.

11. Daniel C. Dennett, *Brainstorms* (Cambridge: Bradford, 1978): 267–85.

12. The primate behaviorist Frans de Waal and four respondents discuss the moral capacities of nonhuman primates in de Waal's *Primates and Philosophers: How Morality Evolved* (Princeton: Princeton University Press, 2006). Consider, in particular, de Waal's discussion of capacities for reciprocity and fairness among humans, chimpanzees, and capuchin monkeys (42–49).

13. See the cartoon "Fair is fair, Larry . . . We're out of food, we drew straws—you lost," depicting three men and a dog, stranded in a life-

boat, with one of the men holding the short straw, in Gary Larson, *The Far Side* (San Francisco: Chronicle Features, 1981).

14. Rawls, *Theory of Justice,* 10–11.

E I G H T   What Kind of Humanity
Do We Embrace?

1. For wide-ranging scholarship on the concept of character in ethics, see Joel Kupperman, *Character* (New York: Oxford University Press, 1995); Christine McKinnon, *Character, Virtue Theories and the Vices* (Calgary: Broadview, 1999); and James Q. Wilson, *On Character* (Washington, D.C.: AEI, 1995). The implications of character for human psychology and health are examined in Christopher Peterson and Martin Seligman, *Character Strengths and Virtues: A Handbook and Classification* (New York: Oxford University Press, 2004).

2. The place for character in environmental ethics is broadly examined in Ronald Sandler, *Character and Environment* (New York: Columbia University Press, 2007); and Ronald Sandler and Philip Cafaro, eds., *Environmental Virtue Ethics* (Lanham, Md.: Rowan and Littlefield, 2005). These studies locate a kindred spirit in Aldo Leopold, in such statements as: "We abuse land when we regard it as a commodity belonging to us. When we see land as a community to which we belong, we may begin to use it with love and respect." Or: "No important change in ethics was ever accomplished without an internal change in our intellectual emphasis, loyalties, affections, and convictions" (Leopold, *A Sand County Almanac: With Other Essays on Conservation from Round River* [New York: Oxford University Press, 1966], x, 225).

3. John Cooper defines *eudaimonia* as "human flourishing" in his land-

mark study *Reason and the Human Good in Aristotle* (Cambridge: Harvard University Press, 1975).

4. James T. Tanner, *The Ivory-Billed Woodpecker* (New York: National Audubon Society, 1942), vii.

5. This is not to say that it is not done, as in the case of some nonrational obedience to self-denial.

6. Gary Beauvais emphasized this for me.